THE MANAGER'S GUIDE TO Statistics and Quantitative Methods

THE MANAGER'S GUIDE TO

Statistics and Quantitative Methods

Donald W. Kroeber

James Madison University

R. Lawrence LaForge

James Madison University

McGraw-Hill Book Company

New York	Johannesburg	Panama
St. Louis	London	Paris
San Francisco	Madrid	São Paulo
Auckland	Mexico	Singapore
Bogotá	Montreal	Sydney
Hamburg	New Delhi	Tokyo

Library of Congress Cataloging in Publication Data
Kroeber, Donald W
The manager's guide to statistics and quantitative methods.

Includes index.
1. Business—Statistical methods. 2. Business mathematics. I. LaForge, R. Lawrence, joint author. II. Title.
HF5691.K78   519.5'02'4658   79-19335
ISBN   0-07-035520-7

1234567890   DODO   89876543210

The editors for this book were William Newton and Joseph Williams, the designer was Mark E. Safran, and the production supervisor was Thomas G. Kowalczyk. It was set in Baskerville by University Graphics, Inc.

Printed and bound by R. R. Donnelley & Sons Company.

TO
OUR
PARENTS

Contents

PART 3.   Quantitative Methods

Appendixes

Preface

Why another book on statistics and quantitative methods for managers? The shelves seem to be full of them, but close examination shows that almost all these books fall into one of two categories, the cute treatment (statistics for people who hate statistics) and the college textbook. The first category tends to support the manager's intuitive feeling that serious quantitative analysis is best left to the professional, and the second category, which was never intended for self-study, confirms it. The result? Another manager who gives up some of the most powerful decision-making tools available to modern management.

Here is a book that attempts to fill the gap between the oversimple approach, with no real managerial application, and the overtheoretical approach, with application lost in the maze of mathematical proofs. The topics are similar to those found in a typical undergraduate program in business administration, but the presentation is oriented for self-study. In selecting and developing the material to be included in the book, the authors had in mind three distinct audiences:

1. Managers whose knowledge of quantitative methods has been gleaned from experience and may therefore be somewhat incomplete or lacking in theoretical foundation

2. Managers who have had an academic background in statistics and quantitative methods but find that time, disuse, and changing methods have limited their ability to apply their rusty skills

3. Men and women who plan to enter an M.B.A. (Master of Business Administration) degree program after an extended absence from col-

lege and feel a need to refresh themselves before taking graduate courses in statistics and quantitative methods.

The four parts of the book can be used in various combinations to meet the differing needs of these three groups. For example, the basic mathematics found in Part 1 should be reviewed by those with no formal mathematical training but may be skipped by others. The statistical methods explained in Part 2 can stand alone as a review for a graduate statistics course or can be used in conjunction with Part 3, Quantitative Methods, as preparation for the minicases presented in Part 4. Some of our colleagues have even suggested that the cases in Part 4 provide an excellent preparation for the quantitative section of M.B.A. comprehensive exams. Other combinations of the four parts can be tailored to the individual reader's needs.

Of particular significance to the occasional user of quantitative analysis are the glossaries in the appendixes, which provide a continuing reference of terms, symbols, and formulas used in statistics and quantitative methods. Appendixes B and C are also cross-referenced to the sections of the book where they are introduced for the reader who requires more than a brief definition.

The objective of producing a serious, comprehensive reference work that can be read and used by nonquantitatively oriented managers often confronted the authors with the dilemma of keeping material simple and understandable yet technically correct. Our response in such instances has been to avoid the "special cases" that often require more complex treatment. Where even common applications require a level of analysis beyond the scope of the book, we have indicated this fact and suggested sources for the advanced or special-interest reader. In several cases where the quantitative analysis is invariably computer-assisted, we completely bypass the actual computational techniques in favor of an explanation of the computer printout.

The authors wish to thank Mrs. Phyllis Price, who typed the draft manuscript; their students at James Madison University, who were (often unwitting) guinea pigs for the pedagogy of the book; and especially their families, without whose patience, understanding, and encouragement this book could never have become a reality.

Donald W. Kroeber
R. Lawrence LaForge

PART 1

Introduction

The models of statistics and quantitative methods are important tools for the modern manager. The two chapters in Part 1 are designed to introduce the concept of modeling and to review some basic mathematical skills that will be needed to work with the models illustrated in later chapters.

Chapter 1 discusses the various types of models and how they are used in decision-making situations. This chapter attempts to set the stage for the statistical and quantitative techniques presented in Parts 2 and 3. Chapter 2, a brief review of selected mathematical procedures and notation, is offered for readers who may be a little rusty.

CHAPTER 1

A Framework for Decision Making

Making decisions is an important part of every manager's job. The successful decision maker relies on such things as experience, judgment, and data made available from reports, financial statements, and other sources. Many of the data available to managers are quantitative; that is, the data are in the form of numbers that must be studied and analyzed in order to arrive at a decision.

It is not surprising that quantitative data are so prevalent in business decision making. After all, most management decisions involve at least some factors that can be *quantified*. For example, decisions regarding inventory levels, sales projections, production schedules, time standards, affirmative action goals, equipment purchases, and many other things involve factors with which we can associate numbers. In any of these situations, the manager's experience and qualitative judgment would be important, but the manager must be equipped to analyze the quantitative data as well.

At least two major advantages result from using quantitative data in business decision making. The first is the ability to communicate information in very precise terms. Numbers simply provide a more precise and unambiguous means of communication than words do. Of course, one must be careful that the numbers are accurate estimates of what they are intended to represent.

A second major advantage is that quantitative data are governed by the basic laws of mathematics and statistics. This allows the decision maker to manipulate the data in order to solve for an unknown variable or to summarize a set of numbers that represent something of importance. For example, it might be useful to have quantitative data concerning the inventory cost of a particular material expressed in the form of a mathematical equation that relates annual cost to the quantity of the material ordered. For possible order quantities under consideration, the equation could be solved using basic algebra for the unknown variable, annual inventory cost.

As another example, quantitative data concerning weekly sales of a particular product over the past year might be expressed in the form of a list or distribution of 52 numbers. Using elementary statistics, the data could be used to compute the average weekly sales, the variation in sales from one week to the next, the range over which weekly sales have varied, and even to test the hypothesis that last year's sales followed essentially the same pattern as those of some previous year.

The ability to analyze data in a quantitative framework often provides important insights into the problem. In a complex situation where many factors are related, it is very difficult, if not impossible, for a manager to conceptualize what will happen as a result of a particular decision. There are just too many factors and interrelationships to keep track of. When quantitative analysis is performed and the related variables are expressed in an equation, the algebraic manipulations used to develop a solution may change the form of the equation but not the basic relationships between the variables. Often the result is that quantitative analysis provides a solution that is not at all intuitively obvious. This fact will be illustrated several times in Parts 2 to 4.

THE USE OF MODELS IN
DECISION MAKING

The mechanism for structuring and analyzing quantitative data in a meaningful way (such as a mathematical equation) is called a *model*. Models of one type or another are used in almost every form of communication and expression. While our primary interest here is with mathematical models useful for business decision making, we first briefly examine the general concept of modeling.

Broadly defined, a model is simply an abstraction of reality that is used to analyze, explain, or predict the real thing it represents. As such, a written report describing a personnel problem is a model. So is a set of blueprints for a new office building, a photograph of the Shenandoah Valley, the Consumer Price Index, and a wind tunnel for testing new aircraft designs.

All these things, while very different, are abstract representations of something. In each case, the model serves a useful purpose. The personnel report provides information to management; the blueprint a technical guide for construction; the photograph a means of remembering a breathtaking view; the Consumer Price Index a measure of purchasing power; and, finally, the wind tunnel a way to test aircraft without endangering the lives of a pilot and crew.

Classifying Models

From the previous examples it is obvious that models are used everyday for a wide variety of purposes. For that reason, it is useful to classify models into broad categories according to their general nature. Figure 1-1 illustrates a widely accepted scheme for classifying models.[1] As the figure shows, every model can be classified as being either physical or symbolic. A physical model is simply one that, in some fashion, physically resembles the real thing of which it is an abstraction. Symbolic models, as the term implies, use symbols such as letters, words, and numbers to represent the thing being modeled.

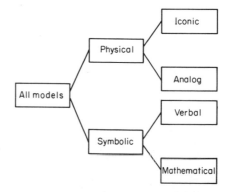

Figure 1-1   Types of models.

    The distinction between symbolic and physical models becomes clearer when one examines the subcategories of each type. A model is physical if it either looks or acts like the thing it represents. The term *iconic* refers to models that *look* like the real thing, while the term *analog* refers to models designed to *act* like the thing they represent. For symbolic models, a distinction is made between models using words and those using mathematical symbols to represent a real-world phenomenon. In Table 1-1, the previous examples of models are classified into the appropriate category.

[1]This classification is suggested by Clifford Springer, Robert Herlihy, and Robert Beggs in *Advanced Methods and Models* (Mathematics for Management Series, vol. 2), Richard D. Irwin, Inc., Chicago, 1965, and Claude McMillan and Richard Gonzalez in *Systems Analysis: A Computer Approach to Decision Models,* 3d ed., Richard D. Irwin, Inc., Chicago, 1973.

TABLE 1-1   *Examples of models*

Category	Type of model	Example
Physical	Iconic	Photograph of Shenandoah Valley; blueprint of new office building
	Analog	Aircraft wind tunnel
Symbolic	Verbal	Personnel report
	Mathematical	Consumer Price Index

Types of Mathematical Models

This book is essentially a practical introduction to the most widely used mathematical models in business decision making. Actually, there are several different types of mathematical models, all having in common the fact that they are developed using the symbols and notation of mathematics. Figure 1-2 provides a technical classification of mathematical models.[2] As shown in the figure, every mathematical model can be classified according to its purpose, the kind of data it uses, and how it works.

In general, the purpose of a mathematical model is either optimization or description. An optimization model is one that is developed to find the "best" solution to a problem. For example, optimization models might be useful to develop solutions to problems expressed like the following:

How much inventory of a particular material should be carried?

What products should be scheduled for production?

How should employees be assigned to tasks?

How much money should be spent on advertising?

Essentially, these are problems that have a best solution, and the mathematical model is designed to identify it. Examples of well-known optimization models include linear programming (Chapter 8) and inventory lot-size models (Chapter 10).

By contrast, a descriptive model merely serves to describe something in mathematical terms. The phenomenon being described might be a single variable or the relationship that exists between several important variables. For example, the following questions might be analyzed with a descriptive model of some type:

How do monthly sales of a particular product vary over the course of the year?

---

[2]Also suggested by Springer, Herlihy, and Beggs, op. cit., and McMillan and Gonzalez, op. cit.

How are sales related to advertising expenditures?

Are college grades a good indicator of management potential?

Are women more likely to be absent from their jobs than men?

The statistical models presented in Part 2 fall into the descriptive category, as well as certain critical-path models (Chapter 9) and waiting-line models (Chapter 11).

In classifying a mathematical model according to the nature of the data it uses, the model is said to be either *deterministic* or *probabilistic*. In a deterministic model, it is assumed that all the data inputs are known with certainty; that is, all the variables that go into the model are known. On the other hand, a probabilistic model is usually more realistic in that it explicitly addresses the problem that one or more of the needed variables may not be known. As the term suggests, probabilistic models associate probabilities with the possible values that an uncertain variable might assume.

As an example of the difference between deterministic and probabilistic models, consider the following simple model for forecasting the revenue generated from the sale of a $10 item:

$$\text{Revenue} = \text{price} \times \text{quantity sold}$$

or

$$\text{Revenue} = \$10 \times \text{quantity sold}$$

If the quantity that will be sold in the forecasted period is known or assumed to be known, the model would be deterministic since a single value could be associated with the variable "quantity sold." On the other hand, if sales are not known, the model would be probabilistic and the variable "quantity sold" would be represented by several possible values, each with its associated probability of occurring. Although more realistic, the probabilistic model is generally more difficult to work with, since we must somehow determine the appropriate value to use in the model.

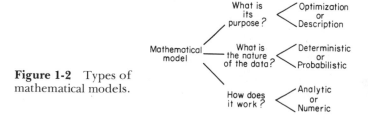

Figure 1-2 Types of mathematical models.

There are many applications of both deterministic and probabilistic models in business decision making.

The third and final classification in Figure 1-2 categorizes mathematical models according to how they work. There is an important distinction between models that are analytic and models that are numeric. A model that works analytically is essentially manipulated or "solved" using standard mathematical or statistical techniques so that a solution is obtained or the model is expressed in some desired form. On the other hand, a model that works numerically is not really solved at all but continually experimented with in order to see what happens under a variety of circumstances.

Sometimes the development of a mathematical model results in such a complex series of equations that it is impossible to find a direct solution. In those situations, the model can be made to work in numeric fashion by asking a series of "what if" questions, using the model to see the likely results. This is essentially a process of repeatedly plugging into the model values for some of the variables in order to see what happens to the others.

Most of the models presented in Parts 2 and 3 fall into the analytic category. However, Chapter 12 deals with a powerful and important numeric technique known as simulation. In that chapter, the distinction between analytic and numeric models is discussed in more detail.

TYPES OF DECISION SITUATIONS

It is also quite useful to classify the types of decision situations faced by managers. Usually the nature of the decision situation determines what type of mathematical model is appropriate. One of the things that makes decision making such a difficult task is that a manager usually does not know what the future holds. Given a crystal ball that describes what is going to happen in some planning period, most managers would make sure that just the right amount of inventory is on hand, the number of new personnel hired and trained is exactly what is needed, the optimal advertising strategy is followed, the most profitable investments are made, and so forth.

The term *states of nature* is sometimes used to represent the possible events which could occur in the future and which will ultimately determine how successful a decision is. The states of nature are defined in such a manner that only one can occur. For example, in making a decision to

·invest in a particular stock, the states of nature might be defined in terms of the value of the stock: it will either go down, stay the same, or go up. Within that framework, it becomes useful to classify decision situations according to the decision maker's knowledge about what state of nature is going to occur.

When the decision maker knows which state of nature is going to occur, the situation is called decision making under *certainty*. When the state of nature is not known, the manager is making a decision under conditions of *risk* or *uncertainty*. It should be pointed out that decision making under certainty is not nearly as simple as it might seem. The term "certainty" in this context does *not* mean that the decision maker knows everything there is to know about the situation. Instead, it implies that he or she knows that a particular event is going to occur and must somehow use that information in deciding what course of action to take. Knowing that the value of a particular stock is going to increase does not necessarily mean that it is the *best* investment one can make with one's money. The difficulty, of course, is that there are many alternative ways to invest money, some of which the decision maker may not even know about.

Decision making under risk or uncertainty is even more complex. In these situations, the decision maker might be able to associate probabilities with each state of nature in order to describe its likelihood of occuring. In the extreme, he or she may have no information at all about the occurrence of each state of nature. When probabilities can be assigned to the states of nature, basic statistical concepts and techniques are useful for analyzing the alternative courses of action being considered. When nothing at all is known about the possible occurrence of each state of nature, the decision can only be based on the degree of optimism or pessimism the decision maker has that a favorable state of nature will occur. Technically, the term *risk* refers to situations where probabilities can be associated with the states of nature and *uncertainty* to situations where no such information is available. Most management decisions are probably made under conditions of risk.

As mentioned earlier; there is a relationship between the nature of the decision situation and the type of mathematical model that is appropriate. Under conditions of certainty, one would expect to find many applications of deterministic models. Also, models that are analytic in nature and designed for the purpose of optimization would seem to be especially compatible with conditions of certainty. On the other hand, decision making under risk, by its very nature, requires probabilistic models. One would also be much more likely to find it necessary to use models that are descriptive in purpose and numeric in nature in a decision situation characterized by risk or uncertainty.

CONCLUSION

Quantitative data and mathematical models play an important role in many decision-making situations. The chapters that follow are designed to illustrate what the basic statistical and mathematical models are all about and how they can be applied to practical management problems. Before introducing statistical models in Part 2, we offer in Chapter 2 a brief review of some basic mathematical concepts needed throughout the book.

CHAPTER 2

Mathematical Concepts in Statistics and Quantitative Methods

Mathematics is undeniably a central force in modern management. Indeed, except for the intimidating effect, this book could just as easily have been titled *Applied Managerial Mathematics*. However, the key word in such a title is *applied*, not mathematics; we are less interested in the mathematical theory of the techniques covered in Parts 2 and 3 than in how those techniques are applied. It is not necessary to understand the theory by which an internal combustion engine operates in order to drive a car, but the car can be used more efficiently with the knowledge of a few basic principles contained in the owner's manual. This chapter is the "owner's manual" to the mathematics of the statistical and quantitative methods that follow.

When you buy your fourth or fifth car, you need not study the owner's manual with the same care exercised on your first car. So it is with this "owner's manual." If you are approaching statistics and quantitative methods for the first time, you should read this chapter carefully. If you are using this book to review techniques you have not used for a while, or if you are accustomed to using other mathematical tools in your work, this chapter can be merely scanned without loss of continuity.

MATHEMATICAL NOTATION

Mathematicians usually try to convey their ideas in clear, precise, and unambiguous terms. Words are often inadequate to meet this task. Even when we can describe a mathematical process clearly with words, it is often

11

inefficient to do so. For example, consider the following two statements: *A sum of money deposited in a savings account at a given annual rate of interest will have added to it at the end of the first compounding period an amount equal to the original sum times the annual rate of interest divided by the number of compounding periods per year. This total becomes the principal for the next compounding period and so on, until the final sum is withdrawn* and

$$S = P(1 + i/m)^{nm}$$

Both statements are symbolic models of the compound interest process. The first, a verbal model, is subject to interpretations of the words used and even to stress given terms when reading the statement. The second statement is a mathematical model, shorter and more precise than the first, but still not clear unless the symbols and notation are known to the reader. In this section, we review the most common mathematical symbols and their uses.

Variables

One of the most powerful tools of mathematical notation is the use of letters to represent terms that vary in numerical value under different conditions. These terms are called *variables* and are identified with symbols according to the fancy of the analyst and certain commonly accepted conventions. For example, the letter i in the compound interest formula represents the annual interest rate, which can take on any value, although we expect it to be in the general range of 5 to 10 percent. Some other variables used in this book are[1]

X, sometimes referred to as a *random variable,* which represents a value that occurs in a probability distribution. X is also used to identify variables plotted on the horizontal axis of a graph.

Y, also used in many cases to represent a random variable. Variables plotted on the vertical axis of a graph are usually designated Y.

μ, the lowercase Greek letter mu, represents the arithmetic mean, or *average.* Greek letters are common in statistics.

S, P, m, n represent the final, or future, sum; the original amount, or principal; the number of compounding periods per year; and the number of years, respectively, in the compound interest formula.

[1]For a complete set of symbols and their meanings see Appendix B.

Subscripted Variables When several observations of the same variable are used in solving a problem, the variable may be *subscripted* to avoid confusion. The subscript itself is a variable and is usually represented with a lowercase i or j. Thus, the height of starters on a basketball team might be designated X_i, where X_1 is the height of the first player, X_2 the height of the second, and so on. When it is obvious that there are many observations and it is not necessary to distinguish between them, the subscript is often dropped.

Subscripts are used in another context to distinguish between like variables. For example, in regression analysis, it is common to designate something called the *dependent variable* as Y and the *independent variable* as X. Since there may be more than one independent variable, they are often designed X_1, X_2, X_3, and so on.

Long series of subscripted variables may be interrupted by *ellipses* (dots) to show the omission of some terms. For example,

$$X_1, X_2, X_3, X_4, X_5, X_6, X_7$$

can be written

$$X_1, X_2, X_3, \ldots, X_7$$

An open-ended series in which the subscript of the last term is not known can be written

$$X_1, X_2, X_3, \ldots, X_n$$

where n refers to the number of terms.

Superscripted Variables The superscript has a very clear, operative meaning in mathematics: it is the exponent or power to which a variable is raised. Thus, X^2 is X to the second power or X times X, usually called X squared. Similarly, X^3 is X to the third power or X times X times X, called X cubed.

The superscript may also be a variable or a term involving several variables, like the term nm in the compound interest formula. In this case, if n is 2 years and interest is compounded quarterly (4 times per year), the exponent is 2 times 4, or 8.

Operative Notation

We are all familiar with the four basic arithmetic operations of addition, subtraction, multiplication, and division as indicated by $+$, $-$, $\times$, and $\div$, respectively. As mathematical operations increase in complexity, however, the notation for these operations may vary slightly.

Addition and Subtraction   Addition and subtraction are noted by the use of the addition or subtraction sign:

$$4 + X + 9\,Y$$
$$27 - Y$$

In some special cases, addition and subtraction are *both* indicated by combining the two signs:

$$15 \pm 4 = 11 \text{ and } 19$$

Summation   The addition of a number of observations of the same variable, say, 10 observations of X, may be written

$$X_1 + X_2 + X_3 + \cdots + X_{10}$$

The same addition can be expressed more conveniently as a *summation:*

$$\sum_{i=1}^{10} X_i$$

which means that the variables X_i are summed (added together) starting with the first X (in which $i = 1$) through the tenth X. In practice, the subscripts usually are dropped and the summation written

$$\sum X$$

Multiplication   The similarity between the multiplication sign, $\times$, and the random variable, X, gives rise to several alternate methods of indicating multiplication. The following expressions illustrate some of the more common methods:

$3X$, 3 times X
$3(X + 2)$, 3 times the sum of X and 2
$(X + 2)(Y + 3)$, the sum of X and 2 times the sum of Y and 3
ΣfX, the short form of $\sum_{i=1}^{n} f_i X_i$, the summation, from 1 to n, of each variable f times the corresponding variable X.

The multiplication takes place *before* the summation:

$$\sum_{i=1}^{n} f_i X_i = f_1 X_1 + f_2 X_2 + \cdots + f_n X_n$$

Division   The division symbol, ÷, is rarely used in higher mathematics. Instead, numbers or variables are written as fractions to show division:

$$X \div 3 = \frac{X}{3} = X/3$$

Care must be exercised to avoid confusing the alternate division symbol, /, with the vertical line, |, which has the meaning "given"

$$X \,|\, Y = \text{the value of } X \text{ } given \text{ a value of } Y$$

Signed Values   In addition to signaling arithmetic operations, the symbols + and − are used to indicate *positive* and *negative* values, respectively. In a certain sense, the sign of a number indicates an arithmetic operation with zero.[2] Thus, +3 can be thought of as zero plus 3 and −5 can be considered zero minus 5. These values are illustrated on the number line shown in Figure 2-1.

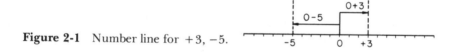

Figure 2-1   Number line for +3, −5.

A few simple rules govern arithmetic operations with signed numbers.

Addition   Numbers (*addends*) of like sign are simply added and given their common sign:

$$(+5) + (+8) = +13 \qquad \text{and} \qquad (-9) + (-6) + (-3) = -18$$

When there are two addends with different signs, the smaller (without regard to sign) is subtracted from the larger (also without regard to sign) and the result is given the sign of the larger:

$$(+5) + (-8) = -3 \qquad \text{and} \qquad (+9) + (-3) = +6$$

---

[2] In practice, plus signs are usually omitted and +3 is written as 3. In this section, when we show plus signs it is for illustrative purposes only.

When more than two numbers with different signs are to be added, the like-signed numbers are added first to give two numbers of different signs which are then added as shown above:

$$(+5) + (-9) + (+8) + (-3) + (-6)$$
$$= [(+5) \ + \ (+8)] + [(-9) \ + (-3) + (-6)]$$
$$= \ (+13) \ + \ (-18) = -5$$

Subtraction   The sign of the number to be subtracted (the *subtrahend*) is changed and the number is added to the number from which it was to be subtracted (the *minuend*):

and
$$(+5) - (+8) = (+5) + (-8) = -3$$
$$(-4) - (-9) = (-4) + (+9) = +5$$

Multiplication   When the signs of two numbers to be multiplied (the *multiplicand* and the *multiplier*) are the same, the result (the *product*) is positive; otherwise, the product is negative:

$$(+5) \times (+8) \ = +40$$
$$(-9) \times \ (-6) = +54$$
$$(-6) \times (+4) \ = -24$$

The multiplication of more than two numbers must be treated as a chain operation with careful attention given to changes in sign:

$$
\begin{array}{rcccccc}
(+5) & \times & (-3) & \times (-2) & \times & (+4) \times (-6) \\
= & (-15) & \times & (-2) & \times & (+4) \times (-6) \\
= & & (+30) & \times & & (+4) \times (-6) \\
= & & & (+120) & \times & (-6) \\
= & & & & -720
\end{array}
$$

Division   The rule for division is similar to the rule for multiplication. When the sign of the number to be divided (the *divisor*) and the sign of the number into which it is to be divided (the *dividend*) are the same, the result (the *quotient*) is positive; otherwise the quotient is negative:

$$(-10) \div (-2) = +5$$
$$(+12) \div (+3) = +4$$
$$(+18) \div (-6) = -3$$

Fractions   Quotients that are not whole numbers (*integers*) are often expressed as *fractions*. The sign of a fraction is determined as in division:

$$\frac{-11}{-3} = (-11) \div (-3) = +\frac{11}{3}$$

$$\frac{+7}{+4} = (+7) \div (+4) = +\frac{7}{4}$$

$$\frac{-9}{+5} = (-9) \div (+5) = -\frac{9}{5}$$

The value of a fraction is unchanged when the upper value (the *numerator*) is either multiplied or divided by the same value as the lower value (the *denominator*):

$$\frac{-8}{+3} = \frac{(-8) \times (+2)}{(+3) \times (+2)} = \frac{-16}{+6} = -\frac{16}{6}$$

$$\frac{-5}{-10} = \frac{(-5) \div (-5)}{(-10) \div (-5)} = \frac{+1}{+2} = +\frac{1}{2}$$

Adding Fractions.   To add fractions, the denominators must be the same. The sum of two or more fractions is found by adding their numerators and retaining the same, *common* denominator of the addends. The rule for multiplying or dividing numerator and denominator by the same value facilitates the conversion to a common denominator:

$$\frac{+11}{-4} + \frac{+9}{+2} = \frac{(+11) \times (-2)}{(-4) \times (-2)} + \frac{(+9) \times (+4)}{(+2) \times (+4)}$$

$$= \frac{-22}{+8} + \frac{+36}{+8} = \frac{+14}{+8} = +\frac{14}{8}$$

When both numerator and denominator can be divided by the same number (and give whole-number quotients), the fraction can be *reduced*. In the case above,

$$\frac{+14}{+8} = \frac{(+14) \div (+2)}{(+8) \div (+2)} = \frac{+7}{+4} = +\frac{7}{4}$$

Subtracting Fractions.   Fractions must also have a common denominator to be subtracted:

$$\frac{+8}{+6} - \frac{-1}{+4} = \frac{(+8) \times (+2)}{(+6) \times (+2)} - \frac{(-1) \times (+3)}{(+4) \times (+3)}$$

$$= \frac{+16}{+12} - \frac{-3}{+12} = \frac{+16}{+12} + \frac{+3}{+12} = \frac{+19}{+12} = +\frac{19}{12}$$

Notice that the sign of the subtrahend is changed by changing *only the sign of its numerator*. To change the signs of both the numerator and denominator leaves the sign of the fraction unchanged.

Decimals   Decimal values are often used as alternatives to fractions for noninteger quotients. Thus,

$$2 \div 8 = \frac{2}{8} = 8\overline{)2.00} = .25$$

Since not all decimals ultimately result in a final whole digit, there is always some question of how to express such a decimal. We use the following rules:

1. A decimal is never more accurate than the number of decimal places in the least accurate number when adding or subtracting.

2. A decimal is never more accurate than the sum of the decimal places of the numbers involved in multiplication or division.

3. Decimals are rounded down when the final digit is 4 or less.

4. Decimals are rounded up when the final digit is 6 or more.

5. Decimals are rounded even when the final digit is 5.

The following examples illustrate these rules:

- Add .334 and .02:

$$.334 + .02 = .354 = .35 \qquad \text{(Rules 1 and 3)}$$

- Subtract .197 from .4567:

$$.4567 - .197 = .2597 = .260 \qquad \text{(Rules 1 and 4)}$$

- Multiply .25 and .35:

$$(.25)(.35) = .0875 \qquad \text{(Rule 2)}$$

- Express .4925 to three decimal places:

$$.492 \qquad \text{(Rule 5)}$$

- Express .875 to two decimal places:

$$.88 \quad \text{(Rule 5)}$$

Of course, these rules may be modified for qualitative reasons. For example, the multiplication of a series of decimals representing probabilities often results in a value to many decimal places more than necessary. Such values are usually rounded to a conventional number of decimal places, such as the four places usually found in probability tables. When rounding more than just the final digit, the rounding rules are modified accordingly. That is, for *two* final digits, 49 is rounded down, 51 up, and 50 even. For *three* final digits, 499 is rounded down, 501 up, and 500 even; and so on.

Exponents and Roots   *Exponents* are superscripts which, as explained earlier, represent the number of times a number or variable is multiplied by itself:

$$X^2 = (X)(X)$$
$$X^3 = (X)(X)(X)$$
$$2^3 = (2)(2)(2) = 8$$
$$(-2)^3 = (-2)(-2)(-2) = -8$$

A *root* is the opposite of an exponent. With the exponent 2 a number is squared or multiplied by itself. The opposite of this process is taking a square root, that is, a value which when multiplied by itself gives the original number

$$\text{Square root of } X = \pm\sqrt{X}$$
$$\text{Square root of } X^2 = \pm\sqrt{X^2} = \pm X$$
$$\text{Square root of } 25 = \pm\sqrt{25} = \pm 5$$

because, in each case, the value on the right multiplied by itself gives the value under the radical sign on the left.

An important feature of the square root is the $\pm$ symbol. For example, $+5$ and -5 are both square roots of 25 because

$$(+5)(+5) = 25 \quad \text{and} \quad (-5)(-5) = 25$$

It also follows from this that there cannot be a square root of a negative number, because no number multiplied by itself will ever yield a negative number if proper treatment of signs is observed.

In many managerial applications, for example, those dealing with numbers of employees, inventory stock levels, and others, a negative root has no real meaning. We therefore drop the ± sign in some later applications, although it is a potentially dangerous habit to acquire.

The following examples illustrate some of the rules for arithmetic operations with values under a radical sign:

$$\sqrt{4} \times \sqrt{6} = \sqrt{4 \times 6} = \sqrt{24}$$
$$\sqrt{8} \div \sqrt{2} = \sqrt{8 \div 2} = \sqrt{4}$$

It is not possible to combine terms under radical signs through addition or subtraction:

$\sqrt{9} - \sqrt{4}$ is not equal to $\sqrt{9 - 4}$ because $3 - 2$ does not equal $\sqrt{5}$

$\sqrt{16} + \sqrt{25}$ is not equal to $\sqrt{16 + 25}$ because $4 + 5$ does not equal $\sqrt{41}$

Roots can also be expressed as *fractional exponents*. For example, the square root can be designated by the exponent ½:

$$\sqrt{X} = X^{1/2}$$

This notation is common when an expression is too large to fit under a standard-size radical sign.

A *negative exponent* indicates that the number or variable is raised to the power of the exponent and then divided into 1

$$X^{-2} = \frac{1}{X^2}$$

The expression X^{-2} is known as the *reciprocal* of X^2, and vice versa, just as ⅓ is the reciprocal of 3.

The widespread availability and use of hand-held electronic calculators has made the computation of square roots virtually a lost art, which we shall not attempt to resurrect here. There are currently available many calculators designed for business, financial, or MBA-student use. All have a square-root capability, and many can do some of the statistical routines described in Part 2. We suggest that the serious reader acquire such a calculator and learn to use it both with this book and on the job.

There are also cube roots, fourth roots, and so on. Their interpretation is analogous to that of the square root.

Nested Operations   Parentheses have been used extensively in this chapter to avoid confusion and to indicate the order in which operations are to be performed. For example,

$$(+3) + (-2) + (+4) \text{ is clearer than } +3 + -2 + +14$$

and $\qquad 3(5 + 2)$ is not the same as $3 \times 5 + 2$

In the second example, the parentheses indicate that

$$3(5 + 2) = 3(7) = 21$$

while $\qquad\qquad 3 \times 5 + 2 = 15 + 2 = 17$

The expression $3(5 + 2)$ can also be interpreted as

$$3(5 + 2) = (3)(5) + (3)(2) = 15 + 6 = 21$$

For more complex operations, expressions sometimes must be nested in a hierarchy of parentheses within brackets within, if necessary, braces. The safest way to perform such nested operations is from the inside out, solving first the expressions within parentheses, then brackets, and finally braces:

$$2\{5[3X(Y + 4) + (5 - 2)^2] + Y(Y - 3)\}$$
$$= 2\{5[3XY + 12X + 3^2] + Y^2 - 3Y\}$$
$$= 2\{15XY + 60X + 45 + Y^2 - 3Y\}$$
$$= 30XY - 120X + 90 + 2Y^2 - 6Y$$

Absolute Values   In some cases, the sign of a value is not important. For example, the difference between milepost 85 and milepost 62 is 23 miles, whether we subtract 62 from 85 and get $+23$ or subtract 85 from 62 and get -23. In such cases, the indication to ignore the sign is given by the *absolute value* symbol $|\ |$

$$|+ X| = |-X| = X$$

and for the milepost example

$$|62 - 85| = |-23| = 23$$

Equalities and Inequalities   The equals sign ($=$) should be familiar to all readers, and we have been using it freely. But sometimes in mathematics it is necessary to point out that expressions are not equal, a relationship indicated by a slash through the equals sign:

$$X \neq 3$$

If additional information is known about the nature of the inequality, we can be more specific. For example,

$X < 3$ means that X is *less than* 3
$X > 3$ means that X is *greater than* 3
$X \leq 3$ means that X is *less than or equal to* 3
$X \geq 3$ means that X is *greater than or equal to* 3

and $1 \leq X \leq 3$ means that X is greater than or equal to 1 (literally, that 1 is less than or equal to X) and also that X is less than or equal to 3. More simply, X lies between 1 and 3, *inclusive*. If 1 and 3 are not possible values of X, that is, X lies between 1 and 3, *exclusive*, we write:

$$1 < X < 3$$

Sometimes a relationship is not known precisely and is estimated or approximated. In such cases, the symbol $\approx$ conveys the meaning "approximately equal to." For example, to show that X is only approximately equal to 3, we write

$$X \approx 3$$

ALGEBRAIC EQUATIONS

We have come virtually a full circle, with a few side trips, back to the mathematical model for compound interest, which we now identify as an *algebraic equation* based on the mathematical notation described in the preceding section. A complete discourse on algebra is not possible here, but there are certain properties of algebraic equations that are important to the understanding of later topics. Of these, the fundamental property is that:

Equality in an algebraic equation is not disturbed by an operation on one side of the equation provided an identical operation is performed on the other side.

Thus, if

$$Y = 7 + 4X$$

an identical quantity can be *added* on both sides of the equal sign:

$$Y + 3 = 7 + 4X + 3$$

Of course, identical quantities can also be *subtracted* without upsetting the equality:

$$Y - 3 = 7 + 4X - 3$$

Both sides can also be *multiplied* by identical quantities *if the multiplication applies to every term.* Thus,

$$Y = 7 + 4X$$
and
$$3Y = 3(7 + 4X)$$
but
$$3Y \neq 3(7) + 4X$$

And the same principle permits each side to be *divided* by identical terms:

$$\frac{Y}{3} = \frac{7 + 4X}{3}$$
or
$$\frac{Y}{3} = \frac{7}{3} + \frac{4X}{3}$$
but
$$\frac{Y}{3} \neq \frac{7}{3} + 4X$$

Finally, both sides of an equation can be *squared* or reduced to their *square roots* without losing the equality:

$$Y^2 = (7 + 4X)^2$$
and
$$\sqrt{Y} = \sqrt{7 + 4X}$$

A second important property is *substitution,* which permits the replacement of one term with an equivalent term. For example, if

$$Y = 7 + 4X \qquad \text{and} \qquad X = 2$$
then
$$Y = 7 + 4(2)$$
$$Y = 7 + 8 = 15$$

The substitution of 2 for X permitted us to *solve* the equation for Y. Substitution combined with other rules facilitates the solution of *any* algebraic equation(s). For example, to solve the following equation for Y:

$$\frac{3Y^2 - 8}{4} = 3X + 4$$

$$\frac{4(3Y^2 - 8)}{4} = 4(3X + 4) \qquad \text{(Multiply both sides by 4)}$$

$$\frac{12Y^2 - 32}{4} = 12X + 16 \qquad \text{(Perform multiplication)}$$

$$3Y^2 - 8 = 12X + 16 \qquad \text{(Perform division on left side)}$$

$$3Y^2 - 8 + 8 = 12X + 16 + 8 \qquad \text{(Add 8 to both sides)}$$

$$3Y^2 = 12X + 24 \qquad \text{(Perform addition)}$$

$$\frac{3Y^2}{3} = \frac{12X + 24}{3} \qquad \text{(Divide both sides by 3)}$$

$$Y^2 = 4X + 8 \qquad \text{(Perform division)}$$

$$\sqrt{Y^2} = \sqrt{4X + 8} \qquad \text{(Take square root of both sides)}$$

$$Y = \sqrt{4X + 8} \qquad \text{(Perform } \sqrt{Y^2}\text{)}$$

Now, if $X = 2$,

$$Y = \sqrt{4(2) + 8} \qquad \text{(Substitution)}$$

and $\qquad Y = \pm\sqrt{16} = \pm 4 \qquad$ (Perform indicated operations)

If this solution seems somewhat tedious, it is because each step is justified algebraically. With practice, many steps can be combined or performed mentally and the process of rearranging an equation becomes fairly routine.

GRAPHING

The process of solving an equation for Y given a value of X does not always promote a full understanding of the relationship between X and Y. In fact, knowing only that $Y = \pm 4$ when $X = 2$ in the equation above is more likely to create an erroneous impression than a valid one. A good technique for demonstrating a two-variable relationship is to plot Y for various values of X on a graph. Such a graph of $(3Y^2 - 8)/4 = 3X + 4$, reduced to $Y = \sqrt{4X + 8}$, is shown in Figure 2-2.

The graph in Figure 2-2 is *nonlinear;* that is, it does not describe a straight line. In managerial analyses, it is more common to assume a *linear* relationship between variables. All linear equations can be reduced to the general form of

$$Y = a + bX$$

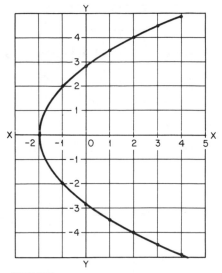

X	Y = √4X+8
-2	0
-1	± 2.00
0	± 2.83
1	± 3.46
2	± 4.00
3	± 4.47
4	± 4.90

Figure 2-2   Graph of $Y = \sqrt{4X + 8}$

where $X, Y =$ variables

   $a = Y$ *intercept* = point where the line crosses, or *intercepts,* the
      Y axis

   $b = $ *slope* of line = change in Y variable for a 1-unit increase in
      X

Figure 2-3 shows the graph of the line

$$Y = 3 + \tfrac{1}{2}X$$

A simple way to plot this or any other straight line is to connect the Y intercept with another point on the line. The Y intercept is plotted easily as +3 on the Y axis. The second point is found by substituting for X. For this example, $X = 8$ gives a Y of $3 + \tfrac{1}{2}(8)$, or $3 + 4 = 7$. The line between $X = 0$, $Y = 3$ and $X = 8$, $Y = 7$ is a *segment* of the line $Y = 3 + \tfrac{1}{2}X$. Of course, in theory, the line itself stretches to infinity in both directions.

   The slope of a straight line is important in several statistical and quantitative methods. As shown in Figure 2-3, the changes in X and Y are designated by the uppercase Greek letter delta, Δ. For this example, a change from point 1 on the graph to point 2 involves a change in X from 2 to 4 and a change in Y from 4 to 5. The slope, or unit change in Y, can be computed as follows:

$$\text{Slope} = \frac{\Delta Y}{\Delta X}$$

$$= \frac{5 - 4}{4 - 2} = \frac{+1}{+2} = +\frac{1}{2}$$

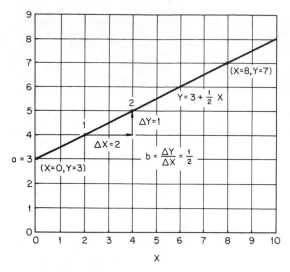

Figure 2-3 Graph of $Y = 3 + \frac{1}{2}X$.

The sign is very important in slope. Here, where an *increase* in Y is associated with an *increase* in X, both changes are positive and the slope is *positive*. Of course, if the change were measured from point 2 to point 1, both ΔX and ΔY would be negative and the slope would still be positive, since slope is determined by the line, not the direction of movement along it. However, when a *decrease* in Y is associated with an *increase* in X, or vice versa, the numerator and denominator will have different signs and the result, the slope, will be *negative*. Lines with various Y intercepts and slopes are shown in Figure 2-4.

SET THEORY

Some relationships between numbers cannot be expressed by algebraic equations or graphs. For example, there is some common, collective relationship between the interest rates of all the commercial banks in the United States. Other managerial data fall into similar logical groupings. In such cases, it is often convenient to describe data in the terminology of *set theory*. Specific applications of set theory to statistical probability are addressed in Chapter 4, but a few definitions are in order here.

Set A collection of related items. For the applications in this book, sets may be considered to consist of related *numbers*, although sets can also consist of tangible objects (a set of dishes, a group of people, and so forth) or intangibles (colors, days of the week, and so forth). In mathematical notation, a set is identified by a listing of its numbers within braces:

The *set* of all possible outcomes from the roll of two dice
$$= \{2,3,4,5,6,7,8,9,10,11,12\}$$

Element   A single item in a set, also referred to as a *data* element. Each of the numbers in the set of rolls of two dice is an element.

Subset   A set within a set, for example,

The *subset* of even rolls of two dice $= \{2,4,6,8,10,12\}$

Elements may be in more than one subset. For example,

The subset of even rolls $= \{2,4,6,8,10,12\}$

and   The subset of rolls divisible by 3 $= \{6,9,12\}$

The subset of rolls divisible by 3 $= \{6,9,12\}$

both contain the elements 6 and 12. Such subsets are considered to be *overlapping*. Other subsets may be *nonoverlapping:*

The subset of even rolls $= \{2,4,6,8,10,12\}$

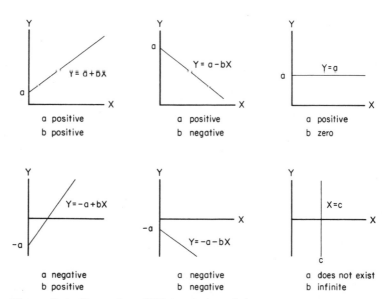

Figure 2-4   Examples of Y intercepts and slopes.

and   The subset of odd rolls = {3,5,7,9,11}

The concept of overlapping and nonoverlapping subsets is particularly useful in explaining certain probabilistic relationships in statistics.

MATHEMATICS AS A LANGUAGE

The example at the beginning of this chapter of the verbal and mathematical models of compound interest is not an exceptional case that overstates the brevity of mathematics compared with conversational English. Indeed, the verbal model probably should be even longer than it is to convey all the information contained in the mathematical model. Brevity is not our aim, however; clarity and understanding are. Consequently, in the chapters that follow, we eschew a mathematical approach in favor of one that is largely verbal. This is not to say that *no* mathematics will be found in subsequent chapters. A certain amount of mathematics is unavoidable and even desirable if one is to use statistics and quantitative methods—desirable because that is the way the professionals in management science and operations research deal with these topics. And if managers, however much they may consider themselves laypeople, are to communicate with these professionals, they must learn at least the basic vocabulary of this language of mathematics.

PART 2
Statistics

The term statistics *has at least three meanings relevant to the material in this book.*

1. Statistics *(plural), in everyday English language usage, refers to a collection of numerical data. Thus, one speaks of the number and dollar value of items sold by members of the sales force as sales statistics, the number of attempts, completions, and yards gained by passing in a football game as passing statistics, and so on. To avoid confusion with other meanings of statistics, we shall use the term* data *(also plural) in such cases.*

2. Statistics *(singular) is also an academic discipline concerned with methods of converting numerical data into information useful for scientific research, business decision making, and other similar purposes. Most readers will be familiar with basic statistical methods for organizing and presenting data in charts, graphs, tables, and summary measures. These techniques are part of* descriptive statistics *and are discussed in Chapter 3. Other methods enable statisticians to draw conclusions about entire populations of data based upon information taken from a sample. These methods are referred to collectively as* inferential statistics. *Inferential methods require some familiarity with statistical theory if they are to be applied properly in decision-making situations.*

29

One fundamental theoretical concept, that of **probability,** *is introduced in Chapter 4. Another important basis of inferential statistics,* **sampling theory,** *is presented in Chapter 5. The inferential processes proper,* **estimation** *and* **hypothesis testing,** *are covered in Chapter 6. A variation of inferential statistics that employs more than one variable,* **regression analysis,** *is introduced in Chapter 7.*

3. *Finally, the term* **statistic** *(singular) is used in a rather narrow sense to describe a summary measure derived from sample data. Of course, when several such measures are discussed, they must be called* **statistics** *(plural) and the possibility of confusion with other meanings is much greater.*

The casual reader may wish to skip Chapter 7 or perhaps leave it to a later time when there is a specific need for the material discussed there. The material in Chapters 3 to 6, however, is essential to all but the most superficial treatment of statistics and is necessary for an understanding of Parts 3 and 4. Of course, if you would like to stay competitive with the new MBA who just joined your organization, you really should read Chapter 7 also.

CHAPTER 3

Descriptive Statistics

Descriptive statistics is primarily concerned with methods of organizing, summarizing, and presenting data. Since business data can be very voluminous, these tasks are often a necessary first step in the business decision-making process. For example, a decision concerning inventory levels cannot be made without information on the demand for each item. This information—drawn, perhaps, from thousands of orders and invoices—must be summarized before it can be used in an inventory decision model. Similarly, a listing of the hourly wage of every worker in a plant may be very accurate and very complete, but it is of little use in comparing that plant's wage structure to industrywide data.

You can probably think of other examples related to your own work, but whatever the type or source of data, the descriptive techniques are similar: data are first collected, then arranged or grouped to reduce volume, perhaps graphed for display, and finally expressed in measures that describe central tendencies and variability. Let us examine these steps in detail.

DATA COLLECTION

Managers today are literally surrounded by data. In fact, a criticism often leveled at computer-based information systems is that they produce *too much* information, much more than management can digest. Yet a great deal of this information is relevant and should be considered in decision making. The first task facing the manager who wishes to collect data, then, is one of screening sources of data with the purpose of eliminating unnec-

31

essary or duplicative data and retaining those which are useful. There are several ways of analyzing sources of data for this purpose.

Internal and External Sources of Data

It is convenient to classify data by their source with respect to a particular organization. Data sources from within the organization are considered *internal* while outside sources are labeled *external*. The relationship of the source to the organization is particularly significant. For example, the logs of miles traveled by United Parcel Service trucks last year is an internal source for UPS but an external source for the Department of Transportation.

Many internal sources of data are maintained as a normal business function. Accounting records, inventory records, payrolls, sales receipts, equipment maintenance records, and expense accounts all fall into this category. Other internal sources may be directed to support a specific information need. Employee surveys and investigations of equipment failure are examples of specified internal sources.

Data from internal sources are usually collected as individual data elements, while data from external sources may be available only in summary form. Consequently, data from internal sources may require more effort to organize and prepare for analysis than is needed for data from external sources. This can be both advantageous and disadvantageous to the user. It is an advantage because individual data elements are subject to broader and often more accurate statistical analysis than summary terms. This difference is demonstrated later in this chapter.

Some managers consider it a disadvantage to work with data from internal sources simply because they do require organization and preparation before they can be used. This should be a disadvantage only to those who feel uncertain or uncomfortable using descriptive statistical methods, a condition we hope to correct in the succeeding sections of this chapter.

Primary and Secondary Sources of Data

Another dichotomous classification of data sources considers the purpose for which data are collected. When a manager collects or has specific data collected to satisfy an information need, we say that the source is *primary*. In contrast, the source of data collected for some other purpose is considered *secondary*, although the data involved may be just as valuable to the proposed analysis as the data from a primary source.

Primary sources may be either internal (observations of a production process, interviews with employees, counts of items in inventory, and so forth) or external (measurements of delivery routes, surveys of customers, tests of competitive products, and so forth). Data from primary sources are generally in the same unprocessed form as internal data and consequently have the advantages and disadvantages associated with internal data.

Secondary sources may also be either internal or external. Internal records and reports maintained to satisfy a legal or contractual requirement, such as profit-and-loss statements, environmental impact statements, accident reports, and employee grievances, are excellent sources of data for statistical and managerial functions as well. These data may not be in the precise form required for a certain statistical method, but they are related directly to the organization and for that reason alone can be of great value. Furthermore, if there appears to be a recurring use for such data, the source department may be persuaded to alter the data format to fit the statistical need.

Data from sources that are both secondary and external are usually the most difficult to use, since they may be both inconveniently summarized and organized for a purpose other than that to which the manager will apply them. While this may be inconvenient, it is often a small price to pay for data that otherwise would be unavailable. Wholesale price indexes, rates of inflation, industry or trade trends, and other similar data can be assembled only with great effort and expense and then only by institutions, such as the federal government, with extensive data collection networks. If a business enterprise needs these types of data, its managers must use them as presented or not at all.

Examples of data from various combinations of sources are shown in Table 3-1.

TABLE 3-1   Examples of data classified by source

Source	Primary	Secondary
Internal	Weights of packages filled by a machine; employee attitudes toward a new policy; length of time a computer program spends in queue; costs of alternative production methods; number of times a copy machine is used daily	Daily demand for an inventory item (from stock records); hourly wage rates (from payroll records); hours of computer time (from meter on computer); overhead costs (from accounting department); transportation costs (from shipping department)
External	Consumer reaction to a new product; number of television advertisements for a competitive product; distances from retail stores to a proposed warehouse location; prices of raw materials required for a production process; shipping time of a supplier	Gross National Product; credit rating of a commercial customer (from Dun and Bradstreet); debt-equity ratio for selected industries (from Dun and Bradstreet); rainfall at a proposed factory site (from the U.S. Weather Service)

TABLE 3-2   Credit-card sales for 1 day

			Good Value Department Store Valleytown, Virginia Credit-Card Sales			
Date: _____						
Cashier: _____						
			Page _____ of _____ pages			
Number	Amount	Number	Amount	Number	Amount	
1	$10.25	15	$21.70	29		
2	25.06	16	6.38	30		
3	14.73	17	15.51	31		
4	8.21	18	12.90	32		
5	19.57	19	20.53	33		
6	38.22	20	3.80	34		
7	20.95	21	24.02	35		
8	23.18	22	45.35	36		
9	35.62	23	14.76	37		
10	13.29	24	31.15	38		
11	33.40	25		39		
12	49.43	26		40		
13	18.33	27		41		
14	58.85	28		42		

ORGANIZING DATA

It was suggested earlier that source documents, such as invoices or sales receipts, are not a form of data convenient for statistical analysis but have to be organized and reduced. The first step in organizing data is simply to extract the relevant data elements from their source and record them, free of extraneous information. Such a listing is referred to as *raw* or *unordered* data. For example, a clerk recording credit-card sales at the end of the day might extract the dollar value of each credit-card sale (ignoring the sales-slip number, the item description, and other information ordinarily found on a sales slip) to produce a list similar to that shown in Table 3-2. This list is unordered with respect to the characteristic being measured, the amount of each sale.[1]

The sales manager who merely glances at the list in Table 3-2 can gain some information from the unordered data. He would know, for example, that:

• The clerk made 24 credit-card sales.

[1] Data unordered with respect to one characteristic may well be ordered with respect to another. These credit-card sales may be in alphabetical order by customers' last name, in numerical order by the stock number of the item sold, or some other unapparent order. With respect to the amount of the sale, however, they are unordered.

- The smallest sale was less than $10.00 but more than $1.00 (there are several three-digit amounts but no two-digit figures).

- The largest sale was less than $100.00 (there are no five-digit amounts).

- Sales in the teens and twenties appear to be more common than sales in higher or lower amounts.

Even these simple facts would be difficult to discern if the data were more voluminous. A computer-prepared listing of all credit-card sales during a month in a large department store would require many pages. In this case, a cursory glance probably would not reveal information about the highest, lowest, and most frequently occurring amounts. We shall not attempt to reproduce a multipage computer printout here, but you can get a rough idea of the difficulty in assimilating large volumes of data by scanning the telephone numbers on a few random pages in your telephone directory and trying to draw some conclusion about the data you find there.

Arrays

The next step in organizing data is to arrange them in numerical sequence. This process is called *ordering,* and the result is an *array,* a list in numerical order. An array may be in either *ascending* or *descending* order, whichever suits the nature or purpose of the data best. Teachers might be more inclined to construct an array of examination grades in descending order to convey the rank or position associated with high grades. Other data might be listed in ascending order to facilitate transfer to a graph where the axes are usually labeled in ascending order. Table 3-3 shows the credit-card sales arranged in an array of ascending order.

TABLE 3-3   Ascending array of credit-card sales for 1 day

Number	Amount	Number	Amount
1	$ 3.80	13	$20.95
2	6.38	14	21.70
3	8.21	15	23.18
4	10.25	16	24.02
5	12.90	17	25.06
6	13.29	18	31.15
7	14.73	19	33.40
8	14.76	20	35.62
9	15.51	21	38.22
10	18.33	22	45.35
11	19.57	23	49.43
12	20.53	24	58.85

The array in Table 3-3 is no briefer than the unordered listing; nonetheless it conveys information more precisely. Now the sales manager draws slightly more specific conclusions from the same brief glance:

- The clerk made 24 credit-card sales.

- The smallest sale (the first item in the array) was $3.80.

- The largest sale (the last item in the array) was $58.85.

- The central values are around $20.

For a small number of data elements, the array is both simple to prepare and easy to interpret. For larger volumes, manual preparation of an array is tedious and error-prone; even computer programs for ordering data are time-consuming (for a computer) and relatively inefficient in the use of storage space. And when a very large array is compiled, the user may find a whole page of $20 sales, rather than the few centrally located ones in our small example. In such cases, data must be condensed to accommodate human limitations of comprehension.

Tally Sheets and Frequency Distributions

The most common method of condensing data is to group them into classes. The actual grouping of data is not difficult and may be done directly from source documents, an unordered data list, or an array. The usual technique is to prepare a *tally sheet* for recording the number of items in each class and then transfer the information to a *frequency distribution* showing the classes and number of items in each class. Although one might not find it necessary to group so few data as the 24 credit-card sales, a tally sheet and frequency distribution for those data have been prepared for illustrative purposes and are shown in Table 3-4.

While the tallying of data is a familiar process for most, it is not immediately clear how the classes are defined for different sets of data. Although there are some mathematical approximations for the appropriate number of classes,[2] it is usually more convenient (and certainly just as valid) to apply subjective criteria. The following "rules" will be helpful in defining the classes in a frequency distribution.

- The number of classes should be between 5 and 20, inclusive.

---

[2]Sturges' rule for the number of classes is $1 + 3.3\log N$, where N is the number of items in the data set. Following Sturges' rule, the 24 credit-card sales would be divided into 5.55 or 6 classes.

- The size of the class, or *class interval,* should be the same for all classes.

- It is convenient, in subsequent statistical computations, for the midpoint, or class mark, of each class to be a whole number.

- The classes must be defined so that every element in the data set falls into one and only one class.

- No class should be empty, i.e., have no items falling in it

TABLE 3-4   Tally sheet and frequency distribution for credit-card sales for 1 day

Tally sheet	
Class	Tally
$ 0.00–$9.99	|||
10.00–19.99	ⅣⅡ |||
20.00–29.99	ⅣⅡ	
30.00–39.99	||||
40.00–49.99	||
50.00–59.99		

Frequency distribution	
Class	Frequency f
$ 0.00–$9.99	3
10.00–19.99	8
20.00–29.99	6
30.00–39.99	4
40.00–49.99	2
50.00–59.99	1

Six classes are selected for the 24 credit-card sales. A value close to the low end of the suggested number of classes is chosen because there are relatively few items in the data set. Specifically, six classes are chosen because six $10 intervals conveniently include all 24 items. The classes are defined as $0.00 to 9.99, $10.00 to 19.99, and so on, because it is common to think of dollars in 5s, 10s, 20s, 100s, etc. Any of these intervals might be appropriate for a given set of dollar values.

   The endpoints, or *boundaries,* of the classes are selected to avoid overlap (a situation in which one number could be in either of two classes) and to ensure that every item falls into some class. For example, if the classes are defined as $0.00 to $10.00, $10.00 to $20.00, and so on, it is not clear into which class a sale of exactly $10.00 should be placed. It is equally inappropriate to leave gaps between classes. If the classes are defined as $0.00 to $9.00, $10.00 to $19.00, and so on, it would be impossible to record a sale of, say, $9.35.

**TABLE 3-5   Ascending array of annual incomes of 35
executives of the ABC Corporation**

$11,500	$17,500	$21,000	$24,500	$30,200
12,700	18,200	21,500	24,800	32,700
13,500	18,700	22,000	25,300	33,500
14,800	19,000	22,000	25,900	37,500
15,200	20,100	22,500	27,000	42,500
16,000	21,000	23,100	28,500	67,500
17,000	21,000	24,000	29,400	85,000

Empty classes can be avoided by redefining the number and/or interval of classes or by using an *open-end class*. For example, consider the annual incomes of 35 executives of the ABC Corporation shown in Table 3-5. These incomes have been grouped into three different frequency distributions shown in Table 3-6.

Table 3-6, Part 1, shows an initial attempt to group income by $10,000 classes, following logic similar to that employed to group the credit-card sales. However, two classes are empty.

A second attempt to group the income data is shown in Part 2. Reducing the number of classes has eliminated the empty classes, but it has also eliminated discrimination between classes: over 90 percent of the data items fall into the first two classes. Besides, there are only four classes and the suggested minimum is five.

The best grouping of these data is shown in Part 3 of Table 3-6, where the relatively small class interval of $5000 provides good discrimination and empty classes are avoided by the open-end class of "$45,000 and over." Open-end classes are expedient when a data set contains a few very high or very low items (or some of each) widely separated from the remaining items. However, while open-end classes are convenient in constructing frequency distributions, they sometimes prove troublesome in other statistical methods.

**TABLE 3-6   Frequency distributions of the salaries of 35 executives of
the ABC Corporation**

(1)		(2)		(3)	
Class	f	Class	f	Class	f
$10,000–$19,999	11	$10,000–$29,999	28	$10,000–$14,999	4
20,000– 29,999	17	30,000– 49,999	5	15,000– 19,999	7
30,000– 39,999	4	50,000– 69,999	1	20,000– 24,999	12
40,000– 49,999	1	70,000– 89,999	1	25,000– 29,999	5
50,000– 59,999	0			30,000– 34,999	3
60,000– 69,999	1			35,000– 39,999	1
70,000– 79,999	0			40,000– 44,999	1
80,000– 89,999	1			45,000 and over	2

More on Frequency Distributions

The frequency distribution showing the *actual number* of items in each class is properly called an *absolute* frequency distribution. Frequency distributions may also show the percentage or *proportion* of items in each class, in which case they are called *relative* frequency distributions. Both absolute and relative frequency distributions can be made *cumulative* by indicating the number (or proportion) of items *in and below* each class or *decumulative* by showing the number (or proportion) of items *in and above* each class. These variations of the frequency distribution are illustrated for the 24 credit-card sales in Table 3-7.

TABLE 3-7   Various forms of frequency distributions for credit-card sales for 1 day

Absolute frequency distribution		Relative frequency distribution		
Class	f	Class		$f*$
$0.00–$9.99	3	$0.00–$9.99	24/3 =	.125
10.00–19.99	8	10.00–19.99	24/8 =	.333
20.00–29.99	6	20.00–29.99	24/6 =	.250
30.00–39.99	4	30.00–39.99	24/4 =	.167
40.00–49.99	2	40.00–49.99	24/2 =	.083
50.00–59.99	1	50.00–59.99	24/1 =	.042

Absolute cumulative frequency distribution		Relative cumulative frequency distribution	
Class	f	Class	f
$0.00–$9.99	3	$ 0.00–$9.99	.125
10.00–19.99	11	10.00–19.99	.458
20.00–29.99	17	20.00 29.99	.708
30.00–39.99	21	30.00–39.99	.875
40.00–49.99	23	40.00–49.99	.958
50.00–59.99	24	50.00–59.99	1.000

Absolute decumulative frequency distribution		Relative decumulative frequency distribution	
Class	f	Class	f
$0.00–$9.99	24	$ 0.00–$ 9.99	1.000
10.00–19.99	21	10.00–19.99	.875
20.00–29.99	13	20.00–29.99	.542
30.00–39.99	7	30.00–39.99	.292
40.00–49.99	3	40.00–49.99	.125
50.00–59.99	1	50.00–59.99	.042

*The division of each frequency by the total number of items ordinarily is not shown in relative frequency distributions and is included here only to illustrate the source of the relative frequencies.

Up to this point, we have considered only *discrete* data; that is, elements in the data set could have only certain specified values. The credit-card sales were always expressed to the nearest whole penny, incomes to the nearest whole dollar, and so on. Some data will always have whole-number values; the number of cars in a parking lot, the number of orders placed by a salesperson, the number of riders in a car pool, and many others. When data are *continuous,* that is, can assume any value between whole numbers, additional care must be exercised in defining classes in a frequency distribution. Two examples of frequency distributions of continuous data are shown in Table 3-8.

TABLE 3-8 Examples of frequency distributions of continuous data

(a) Age of employees in a small business		
Class, years	Absolute f	Cumulative absolute f
Under 20	3	3
20 and under 25	8	11
25 and under 30	12	23
30 and under 35	15	38
35 and under 40	24	62
40 and under 45	16	78
45 and under 50	5	83
50 and over	10	93
(b) Weight of 90 hams in a shipment		
Weight, pounds	Relative f	Decumulative relative f*
5.0 and under 5.4	.044	.999
4.6 and under 5.0	.144	.955
4.2 and under 4.6	.244	.811
3.8 and under 4.2	.289	.567
3.4 and under 3.8	.189	.278
3.0 and under 3.4	.056	.089
Under 3.0	.033	.033

*Relative frequencies do not total 1.000 due to rounding.

One common, convenient method for dealing with continuous data is to use an "____and under____" definition of classes. Then, no matter how accurately one wishes to measure age, weight, height, or some other continuous variable, there will never be a problem of an item falling in a gap between classes or knowing which class it belongs in. For example, in part (a) of Table 3-8, an age of 24 years, 11 months, and 29 days in decimal form is expressed as 24.997260274 . . .years. Instead of defining classes to three or four decimal places, we have used "20 and under 25," "25 and under 30," and so on. Since 24 years, 11 months, and 29 days is under 25 years, it clearly falls into the "20 and under 25" class. It is equally unambiguous that an age of exactly 25 years falls in the "25 and under 30" class. Although the number 25 appears in two class definitions, there is no overlap because of the "and under" term.

The frequency distributions in Table 3-8 also illustrate several previously discussed characteristics. Part (a) shows an ascending absolute and cumulative absolute frequency distribution open at both ends, apparently to accommodate a few very old and very young employees. Part (b) shows a descending relative and decumulative relative frequency distribution with one open-end class.

GRAPHICAL PRESENTATIONS

If, as often asserted, a picture is worth a thousand words, a graph ought to be worth, at least, a few hundred numbers. The comparison is not as frivolous as it may seem. Many individuals, including some in management who deal with numerical data on a daily basis, feel uncomfortable when placed in a position that requires analysis or interpretation of numbers. Managers with this shortcoming must rely on the work of subordinates or, worse, ignore numerical data and base their decisions solely on qualitative information. For such managers, graphical presentations of quantitative data are usually far less threatening and much more likely to be incorporated into the decision-making process. Several easy-to-understand graphical techniques follow logically from the frequency distribution.

Histograms

A histogram is literally a picture of a frequency distribution. On one axis of a graph, usually the horizontal, or X, axis, are displayed the boundaries of the classes while the other axis, the vertical, or Y, axis, represents the frequency. The histogram itself consists of rectangles formed by the class boundaries, the horizontal axis, and lines corresponding to the class frequencies. A histogram of the credit-card sales is shown in Figure 3-1.

Histograms are always drawn without gaps between class rectangles. Theoretically, the definition of classes of discrete data suggests that there should be a slight gap—.01 (1 cent) in the credit-card sales example—between adjacent rectangles. In practice, this is never done, and the result is a clearer, easier-to-read graph.

Histograms drawn from frequency distributions with "____and under____" class definitions have unambiguous class boundaries, since the end of one class is adjacent to (or, at most, an infinitely small distance away from) the next class. Open-end classes present a problem, however,

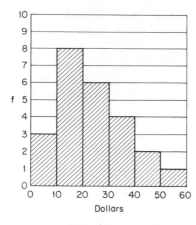

Figure 3-1 Histogram of credit-card sales for 1 day.

since they cannot be represented by a closed rectangle. A "___and under___" frequency distribution and the corresponding histogram are shown in Figure 3-2.

Notice that the "under 3" class in Figure 3-2 is not an open-end class, as it might first appear to be. "Under 3" is the same as "0 and under 3," which gives the first class the same interval as the other classes.

Histograms may also reflect the relative and cumulative aspects of frequency distributions. Of course, a histogram based on a relative frequency distribution will look exactly like one based on an absolute frequency distribution except that the vertical axis will be labeled differently. The num-

Class (years)	f	
Under 3	20	(.187)
3 and under 6	24	(.224)
6 and under 9	18	(.168)
9 and under 12	13	(.121)
12 and under 15	9	(.084)
15 and under 18	5	(.047)
18 and under 21	4	(.037)
21 and under 24	6	(.056)
24 and under 27	3	(.028)
27 and under 30	2	(.019)
30 and under 33	2	(.019)
33 and under 36	1	(.009)

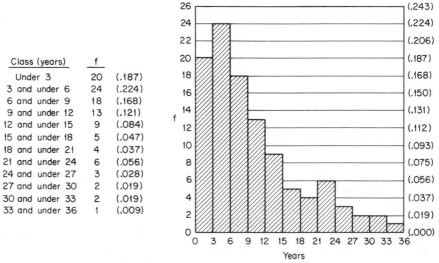

Figure 3-2 Length of employment of workers in a garment factory, absolute and relative.

bers in parentheses to the right of the frequency distribution and on the right side of the histogram in Figure 3-2 show the labeling for a relative frequency distribution and its corresponding histogram.

Cumulative frequency distributions have histograms that characteristically slope upward, while the histograms of decumulative frequency distributions slope downward. Figure 3-3a shows a histogram based on an absolute cumulative frequency distribution and Figure 3-3b shows one based on a relative decumulative frequency distribution

The choice of the type of histogram to use depends on the questions the histogram is intended to answer. An absolute histogram best answers

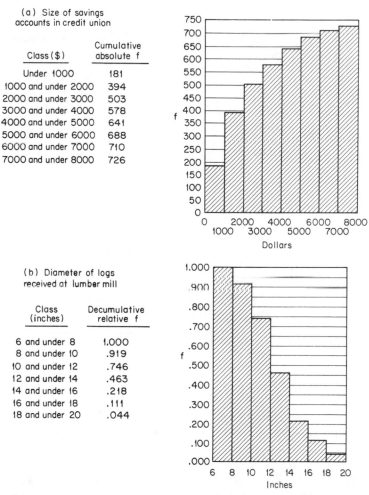

(a) Size of savings accounts in credit union

Class ($)	Cumulative absolute f
Under 1000	181
1000 and under 2000	394
2000 and under 3000	503
3000 and under 4000	578
4000 and under 5000	641
5000 and under 6000	688
6000 and under 7000	710
7000 and under 8000	726

(b) Diameter of logs received at lumber mill

Class (inches)	Decumulative relative f
6 and under 8	1.000
8 and under 10	.919
10 and under 12	.746
12 and under 14	.463
14 and under 16	.218
16 and under 18	.111
18 and under 20	.044

Figure 3-3 Examples of relative and cumulative/decumulative histograms.

the question: How many items are in each class? A histogram based on a relative frequency distribution gives the proportion or percentage of the total in each class. A manager more interested in the number of items less than or more than some class boundary should use a cumulative absolute or decumulative absolute histogram, respectively. Finally, if it is the proportion or percentage of items less than or greater than some class boundary that is desired, cumulative relative or decumulative relative histograms are appropriate. The following questions and answers, based on the data illustrated in Figures 3-2 and 3-3, demonstrate these points.

- How many employees of the garment factory have between 12 and 15 years of employment? (9)

- What percentage of employees have worked at the garment factory between 3 and 6 years? (.224, or 22.4 percent)

- How many credit union accounts have less than $1000? (181)

- How many credit union accounts have less than $5000? (641)

- What proportion of logs are 10 inches or more in diameter? (.746)

- What percentage of logs are 14 inches or more in diameter? (21.8 percent)

Bar Graphs

Another common graphical representation of frequency distribution data is the *bar graph*. A bar graph can be thought of as a histogram with a line or bar representing frequency instead of a rectangle. The bar originates at the midpoint, or *class mark*, of a class. The class mark, which will be used in several other applications, can be found by the formula

$$\text{Class mark} = \frac{\text{lower boundary} + \text{upper boundary}}{2}$$

A frequency distribution and its corresponding bar graph are shown in Figure 3-4.

While the choice between a histogram and a bar graph is sometimes nothing more than a matter of personal preference, there is at least one advantage to using a bar graph for discrete data. When classes are defined in whole numbers, the problem of closing the gap between classes does not arise. In the credit-card sales example, the gap between, say, $9.99 and $10.00 is practically impossible to discern on a scale of 0 to 60 dollars. In that case, $10.00 was used as both the upper boundary of the first class and the lower boundary of the second class. Now, in the taxi-fare example

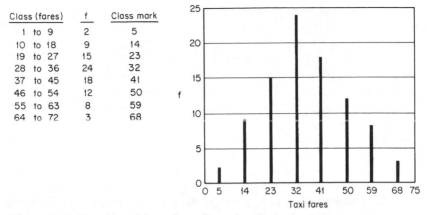

Class (fares)	f	Class mark
1 to 9	2	5
10 to 18	9	14
19 to 27	15	23
28 to 36	24	32
37 to 45	18	41
46 to 54	12	50
55 to 63	8	59
64 to 72	3	68

Figure 3-4 Number of fares for a fleet of taxi cabs for 1 day.

in Figure 3-4, the difference between 9 and 10 fares is not difficult to measure on an accurately drawn graph and 10 should not be used arbitrarily as a common class boundary. If a histogram must be drawn in such a case, the best alternative is to "split the difference" and use class boundaries of 9.5, 18.5, 27.5, and so on. However, this is not really a good solution, since a value of 9.5 (9½ fares) is obviously an impossible daily total for a taxi. The bar graph avoids this problem by using a single line at the class mark.

The advantage in selecting a number of classes and a class interval that give a whole number as a class mark is also illustrated in the taxi-fare example. When a class interval of nine has been determined (there are nine possible values in each class) and the first class mark has been computed as five, subsequent class marks can be obtained by repetitive addition of the class interval:

Previous class mark		Class interval		Next class mark
5	+	9	=	14
14	+	9	=	23
. . .	. . .	. . .	. . .	. . .
59	+	9	=	68

Frequency Polygons and Frequency Curves

In the bar graph, data plots appeared at the end of lines or bars extending from the class marks. When adjacent plots are connected with straight lines (and the bars omitted), the graphical representation is called a *frequency polygon*. A *frequency curve* is a variation of a frequency polygon in which the data plots are fitted to a smooth curve. An example of each and the frequency distribution and bar graph from which they are derived are shown in Figure 3-5.

Although frequency polygons and curves are based upon bar-graph data, they are more similar to histograms in that they reflect frequency as an *area* rather than a *length*. Thus, the frequency polygon and curve take advantage of the ability of the human eye to assimilate information conveyed by area better than information conveyed by length. To illustrate this point, try to visualize, on each of the three graphical representations in Figure 3-5, a line that separates the upper half of assembly times from the lower half. Most people will come close to the broken line between 22 and 23 seconds on the frequency polygon and curve but will have difficulty in finding a similar line on the bar graph. Part of this difficulty is due to the discontinuous nature of the line segments that make up the

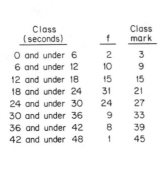

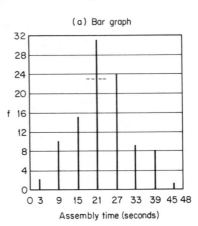

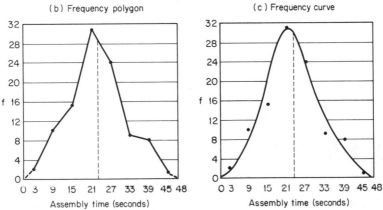

Figure 3-5   Times required to assemble toy trucks (broken lines separate times in the upper half from times in the lower half).

bar graph and part to the inherent difficulty in interpreting length. Furthermore, it can be seen that every point on a bar graph must be read as one of the class marks and the division between upper and lower halves there is almost meaningless.[3]

Ogives

An even better way to estimate points that correspond to certain percentages is based on curves constructed from cumulative or decumulative frequency distributions. Such curves are called *cumulative ogives* or *decumulative ogives*, depending on the type of frequency distribution from which they are drawn. If, as suggested, the purpose of an ogive is to determine percentages, it should be based on a relative frequency distribution. However, one can just as easily construct an ogive from an absolute frequency distribution and obtain values in numbers of items rather than percentages or proportions. Figure 3-6 shows an absolute decumulative ogive and a relative cumulative ogive, both based on the same toy-truck assembly times used in the frequency-polygon (frequency-curve) example.

Because the decumulative ogive in Figure 3-6a shows the proportion of assembly times equal to or greater than a specified time, it is necessary to plot the decumulative frequency for each class at a point that includes the items in that class. The point that satisfies this requirement is the class lower boundary. Similarly, the cumulative ogive in Figure 3-6b shows the number of times equal to or less than a specified time and has cumulative frequencies plotted at the class upper boundary.

The line that separates the upper and lower halves of the distribution of toy-truck assembly times, which had to be visually estimated on the frequency curve, can be read directly from the ogives. The broken line in Figure 3-6a shows that the time associated with the 50th assembly (in an array) is between 22 and 23 seconds. Likewise, the division between the upper and lower 50 percent of the distribution is shown by the broken line between 22 and 23 seconds in Figure 3-6b. The assembly times associated with other positions or percentages can be found similarly.

MEASURES OF CENTRAL TENDENCY

If the purpose of reducing a data set is merely one of eliminating bulk or facilitating graphical presentation, the frequency distribution is an excellent technique, but there are times when even the seven or eight classes of a frequency distribution are too many. A frequency distribution is incon-

---

[3]For the curious, the line between the upper and lower halves on the bar graph was determined by counting off 50 (one-half of 100) assembly times, 2 in the first class, 10 in the second, 15 in the third, and 23 in the fourth, and drawing the line at that point.

(a) Decumulative absolute ogive

Class (seconds)	Decumulative absolute f
0 and under 6	100
6 and under 12	98
12 and under 18	88
18 and under 24	73
24 and under 30	42
30 and under 36	18
36 and under 42	9
42 and under 48	1

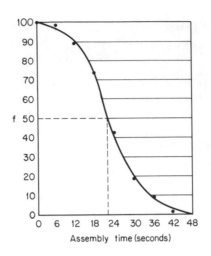

(b) Cumulative relative ogive

Class (seconds)	Cumulative relative f
0 and under 6	.02
6 and under 12	.12
12 and under 18	.27
18 and under 24	.58
24 and under 30	.82
30 and under 36	.91
36 and under 42	.99
42 and under 48	1.00

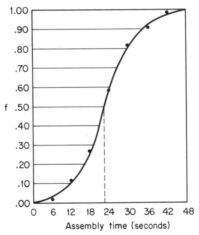

Figure 3-6 Ogives of times required to assemble toy trucks.

venient to use in a mathematical equation and, at best, awkward in everyday speech. In these cases, it is much better to use a single value to describe the data set.

Single values that describe a data set are called *measures of central tendency* because they are usually near the center of an array and there is an inclination or tendency for all elements of the set to behave as if each had that particular value. Measures of central tendency are sometimes called *averages*, although that term is commonly associated with one particular measure, the arithmetic mean. Because of possible confusion between the specific and generic definitions, we shall avoid the use of "average" in discussions of statistical methods.

Populations and Samples

Before discussing individual measures of central tendency, we must specify the nature of the data set represented by those measures. If the set contains all possible items of interest, we call it a *population*. On the other hand, if the set contains only a few random or otherwise representative elements, we call it a *sample*. Measures of central tendency and other characteristics that describe a population are called *parameters*, while the corresponding measures and characteristics of a sample are called *statistics*.

The distinction between a sample and a population often depends on one's perspective. The expiration dates on cartons of milk in a grocery store represent a population to the store manager, but they may constitute a sample (or part of a sample) for the city health inspector.

There are also slight differences of notation and computation between population parameters and their corresponding sample statistics. Until one is familiar with the application and interpretation of various measures, these differences are often confusing. Therefore, in this chapter, all data sets are considered to be populations and accordingly all measures of central tendency and variability are parameters. Samples and statistics are addressed in Chapter 5.

The Arithmetic Mean

The most common measure of central tendency is the *arithmetic mean* or, simply, the *mean*.[4] The mean is, as almost everyone knows, equal to the sum of all the elements in a set divided by the number of elements. In mathematical notation we write

$$\mu = \frac{\Sigma X}{N} \tag{3-1}$$

where μ = mean of population
ΣX = sum of all elements in population
N = number of elements in population

To compute the mean credit-card sales from the data in Table 3-2, simply total the individual amounts and divide by the number of sales

$$\mu = \frac{\Sigma X}{N} = \frac{10.25 + 25.06 + 14.73 + \cdots + 31.15}{24} = \frac{565.19}{24} = \$23.55$$

---

[4]There are other means, such as the geometric mean and the harmonic mean, but they are beyond the scope of this text and are not addressed here.

The mean has a number of interesting characteristics that should be considered before using it to describe a data set.

Every element in the data set contributes to the value of the mean. The mean is the only common measure of central tendency with this somewhat democratic quality.

The mean is influenced by extreme values. This is the price one pays for including every element. The example is often used, apocryphally, of a tiny sheikdom in which the sheik's income is enormous (say, $50 million per year) while the populace receives a pittance (say, $200 per year, each). If there are 20,000 in the populace and only one sheik, the mean income in the sheikdom is still $54,000,000/20,001, or almost $2700 per year! Some researchers, particularly when dealing with small data sets, will "throw out" extreme values to avoid distortion of the mean. This must be done only with great care and only when one is so thoroughly familiar with the data that there is no doubt of what constitutes an "extreme value." A better solution is to use a different measure of central tendency.

The mean may not be an actual element of the data set. There is, for example, no credit-card sale of exactly $23.55 in Table 3-2. This is of little consequence when dealing with dollars, but if the data set consists of dress sizes, a nonexistent value could be confusing or even misleading. For example, a mean dress size of 7 is not representative of a population consisting mainly of 6s and 8s, since odd dress sizes are cut for a different figure than even sizes. Again, a different measure of central tendency may be called for.

The mean is important in further statistical computations. This importance is based on two mathematical properties of the mean: the sum of deviations from the mean is zero and the sum of squared deviations is minimized when those deviations are measured from the mean. In mathematical notation, we express these properties as

$$\Sigma(X - \mu) = 0$$
and
$$\Sigma(X - \mu)^2 = \text{minimum}$$

Neither of these properties argues for selection of the mean over some other measure of central tendency, but they tend to point out the fundamental role of the mean in statistics.

The Median

The *median* is usually defined as the central value in an array. This definition may be slightly misleading, since it implies that the median actually occurs in a data set. In reality, the median is certain to exist as an actual data element only when there is an odd number of elements in the set.

For example, the central value in an array of 25 numbers is the 13th value, since there are 12 values higher and 12 lower. However, if the set contains only 24 elements, we must create an artificial position, the 12.5 position, to obtain equal numbers of higher and lower values. To generalize, we can write an expression for the median position as

$$\text{Median position} = \frac{N + 1}{2}$$

It should be emphasized that this formula gives the value of the median *position,* not the actual value of the median. To find the true median value one must refer back to the array. If the median position is a whole number, one simply counts down to that position and notes the value of the element found there. When the median position is not a whole number, the median is considered to be the mean of the values immediately preceding and following the median position. For the 24 credit-card sales arrayed in Table 3-3, the median is found as follows:

$$\text{Median position} = \frac{N + 1}{2} = \frac{24 + 1}{2} = \frac{25}{2} = 12.5$$

12th position = 20.53
13th position = 20.95

$$\text{Median} = \text{mean of } 20.53 \text{ and } 20.95 = \frac{20.53 + 20.95}{2.} = \frac{41.48}{2} = \$20.74$$

The median also has a number of interesting properties that should be considered when selecting an appropriate measure of central tendency.

The median is not influenced by extreme values. Sometimes only one and at most only two elements determine the value of the median. This makes the median particularly attractive when describing the central tendency of data that are limited in one direction only. Income data illustrate this point well: there is a definite lower limit to income (zero, if one disregards the possiblity of a net deficit) while the upper limit is virtually unbounded. The median of $200 tells much more about the central tendency of income in the hypothetical sheikdom than the mean of $2700 does.

The median may not be an actual element of the data set. The median is not an actual data element in the approximately one-half of all sets that contain an odd number of elements. This property has the same significance for the median as it has for the mean.

The median is not used in any common subsequent statistical methods. We ordinarily do not speak of "deviations from the median" or otherwise use the median in higher-order statistics.

The median divides a frequency curve into equal areas. The broken line that separates the upper and lower halves of toy-truck assembly times in Figure 3-5c represents the median time of that distribution.

The Mode

The *mode* is quite simply the most frequently occurring value in a data set. The mode is more appropriate for distribution of data that are measured only in discrete sizes, quantities, or other dimensions. Table 3-9 shows an array of the sizes of metric wrenches used in a machine shop during an 8-hour period. The single value most representative of the central tendency of these data is not the mean (12.083 millimeters) or the median (11 millimeters), but the mode, 10 millimeters, which occurs 13 times, more than twice as often as any other value. Thus, if the shop manager wishes to stock additional tools, the first choice among wrenches should be the 10-millimeter size because demand for this size is the greatest and a shortage of 10-millimeter wrenches is most likely to cause a work stoppage.

The mode, too, has certain properties of interest.

A data set may have more than one mode. When two values occur an equal number of times (more than once and more than any other value), we say that the set is *bimodal*. A data set may also have three or more modes, but in such cases the worth of the mode as a measure of central tendency is questionable.

TABLE 3-9   Ascending array of metric wrenches used in a machine shop during an 8-hour period

No.	Size, millimeters	No.	Size, millimeters	No.	Size, millimeters
1	6	17	10	33	13
2	6	18	10	34	14
3	7	19	10	35	14
4	8	20	10	36	14
5	8	21	10	37	14
6	8	22	10	38	15
7	8	23	10	39	16
8	9	24	11	40	16
9	9	25	11	41	16
10	9	26	12	42	16
11	10	27	12	43	18
12	10	28	12	44	18
13	10	29	12	45	18
14	10	30	12	46	20
15	10	31	13	47	20
16	10	32	13	48	22

A data set may have no mode. If no value in a data set occurs more than once, we say that the set has no mode. Sets of continuous data, such as measurements of length, weight, or some similar dimension, characteristically have no mode. The theory underlying this fact is that additional precision (another few decimal places, perhaps) will always reveal some difference between seemingly identical measurements of continuous data.

When there is a mode, it will always be an element in the data set. This property is obvious from the definition of the mode.

The mode has no common use in further statistical computations. The mode is similar to the median in this respect.

The highest point on a frequency curve occurs at the mode. If one thinks of a frequency curve as a smoothed histogram in which the class interval is 1, the highest rectangle (and consequently the highest point on the frequency curve) will be at the mode. Obviously there are limitations to this interpretation, since we know that these may not be a mode and the histogram analogy is certainly not valid for continuous data. Nevertheless it is often convenient to think of the mode in such terms.

Computing Measures of Central Tendency from Grouped Data

Up to this point, we have worked exclusively with individual data elements to determine the three measures of central tendency. All three can also be approximated by computational methods when data are grouped in a frequency distribution.

The Computational Mean   The formula for the computational mean is

$$\mu = \frac{\Sigma f(CM)}{N} \tag{3-2}$$

where f = frequency of each class

   CM = class mark, or midpoint, of each class

   N = number of elements in the population

The mean should be computed from the individual data elements if they are available. However, for illustrative purposes, the mean of the credit-card sales is computed from grouped data in Table 3-10. The computational mean of $23.75 determined in Table 3-10 is not the same as the true mean of $23.55. Slight differences between the true and computational mean are not unusual and tend to become smaller with larger data sets.

The formula for the computational mean easily can be adapted to one for a *weighted mean*. A weighted mean is appropriate when elements in a

TABLE 3-10   Computational mean of 24 credit-card sales for 1 day

Class	f	CM*	f(CM)
$ 0.00–$9.99	3	$ 5.00	15.00
10.00–19.99	8	15.00	120.00
20.00–29.99	6	25.00	150.00
30.00–39.99	4	35.00	140.00
40.00–49.99	2	45.00	90.00
50.00–59.99	1	55.00	55.00
	$N = 24$		$\Sigma f(CM) = 570.00$

$$\mu = \frac{\Sigma f(CM)}{N} = \frac{570.00}{24} = \$23.75$$

*Rounded to the nearest whole $0.01. Selection of classes to give a whole-number class mark simplifies this computation.

data set are given unequal emphasis or weights. For example, to find the weighted mean interest of the four savings accounts described in Table 3-11, one uses the formula

$$\mu = \frac{\Sigma wX}{\Sigma w} \tag{3-3}$$

where w is the weight assigned each X. In the interest problem, interest rates are weighted by the account size.

TABLE 3-11   Weighted mean interest for four savings accounts

Type of account	Size of account, w	Interest rate, X	wX
Passbook	500.00	.0525	26.25
One-year	1000.00	.0600	60.00
Two-year	1500.00	.0650	97.50
Three-year	3000.00	.0700	210.00
	$\Sigma w = 6000.00$		$\Sigma wX = 393.75$

$$\mu = \frac{\Sigma wX}{\Sigma w} = \frac{393.75}{6000.00} = .065625$$

Observe that the weighted mean interest of .065625 (approximately 6.56 percent) is not the same as the simple mean obtained by summing the four rates and dividing by the number of rates, which gives .2425/4 or

.060625 (approximately 6.06 percent). In this example, use of the simple mean would understate the annual interest earnings by $30.00:

Earnings with weighted mean: 6000 × .065625 = $393.75
− Earnings with simple mean: 6000 × .060625 =   363.75
 Difference $ 30.00

The Computational Median   An approximation of the median can be computed from a frequency distribution by the formula

$$\text{Median} = L + \text{CI} \left(\frac{N/2 - a}{b} \right) \tag{3-4}$$

where L = lower boundary of the median class
 $\quad$ CI = class interval
 $\quad$ N = number of elements in the population
 $\quad$ a = number of elements in classes below the median class
 $\quad$ b = number of elements in the median class

The formula for the computational median seems somewhat imposing at first, but the principle embodied is quite simple. The computational median is proportional to the position of the median in the median class. For the 24 credit-card sales illustrated here, the median is one-sixth of a class interval into the class that contains the true median position. When the fractional class interval is added to the lower boundary of the median class, the result is the computational median of $21.67. Once again, we see that a measure of central tendency computed from a frequency distribution is not necessarily equal to the true measure, in this case a median of $20.74.

$$\begin{aligned}
\text{Median} &= L + \text{CI} \left(\frac{N/2 - a}{b} \right) \\
&= 20.00 + 10.00 \left(\frac{12 - 11}{6} \right) \\
&= 20.00 + 1.67 = \$21.67
\end{aligned}$$

Earlier, we suggested that the median divided a frequency curve into equal areas and graphically determined such a line for a distribution of 100 toy-truck assembly times in Figure 3-5. We can now confirm that estimate of "between 22 and 23 seconds" with the computational median:

$$\text{Median position} = \frac{N + 1}{2} = \frac{100 + 1}{2} = \frac{101}{2} = 50.5$$

Lower boundary of class containing 50.5 position: $L = 18$
Class interval: $CI = 6$
Number of elements in classes below median class: $a = 27$
Number of elements in median class: $b = 31$

$$\text{Median} = L + CI\left(\frac{N/2 - a}{b}\right) = 18 + 6\left(\frac{50 - 27}{31}\right)$$
$$= 18 + 4.45 = 22.45 \text{ seconds}$$

The Computational Mode   There are several ways to compute a mode from a frequency distribution; however, a computational mode is something of a contradiction in terms. Since it is computed, it will not necessarily be a value in the data set, yet the mode is supposed to be the most frequently occurring value. Furthermore, computational methods will not reveal second or third modes, if they exist. For these reasons, we shall not attempt a computational mode but instead suggest two alternative methods of expressing modality in a frequency distribution, the *modal class* and the *crude mode*.

The modal class is the class with the greatest frequency. In the credit-card sales example, the modal class is $10 to $19.99 with a frequency of 8.

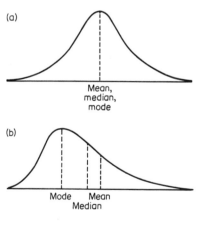

(a)

Mean,
median,
mode

(b)

Mode   Mean
Median

(c)

Mode 1 Mean, Mode 2
median

Figure 3-7 Mean, median, and mode for three different data sets.

There can, of course, be more than one modal class in a frequency distribution.

The crude mode is the class mark of the modal class. For the credit-card sales example, the crude mode is $15.00. Obviously, if there can be more than one modal class, there can be a like number of crude modes.

Choice of an Appropriate Measure of Central Tendency

We have already suggested that the median is more appropriate than the mean when dealing with data that are limited in only one direction and that the mode is appropriate for data that are measured in sizes. Let us look more closely at the question of which measure of central tendency to use to describe different categories of data.

Figure 3-7 shows frequency curves for three different data sets. Figure 3-7a represents a symmetrical distribution in which the mean, median, and mode all occur at the same point, a characteristic of a *normal distribution* (Chapter 4). In this case, it might seem that the choice is immaterial, since any of the three measures of central tendency will give the same result. While this is true, we would still give preference to the mean because of its greater value in subsequent statistical computations.

Figure 3-7b is a *skewed* frequency curve; that is, one tail is elongated. This particular curve exhibits *positive skewness* or is *skewed to the right*. A negatively or left-skewed curve would have an elongated left tail. For skewed data, we would be more inclined to select the median to describe the data set. The skewed curve also illustrates an interesting empirical relationship between the mean, median, and mode: for nonsymmetrical curves, the median falls between the mean and the mode, the distance between the median and mode being approximately twice that between the median and the mean.

Figure 3-7c is also a symmetrical curve, but it is bimodal. Although the mean and the median are centrally located, they are not the most representative measures of the data set. The best measure in this case is the mode, or rather the *two* modes.

MEASURES OF VARIABILITY

A sauna bather who alternately runs naked through the snow and sits perspiring in a steam-filled room experiences a mean temperature of, perhaps, 80 degrees Fahrenheit. So does a person dozing in a hammock on a warm summer afternoon. Clearly, there is more to describing a data set (in this case, a set of temperatures) than merely expressing a measure of central tendency. The *variability* of data is often just as important as the measure of central tendency. Let us look at two ways, one simple and one not so simple, of expressing variability.

The Range

The *range* of a data set is the difference between the highest and lowest values in the set. When data are arrayed in ascending order, the range is simply the value of the last element minus the value of the first.

The range of the 24 credit-card sales is $58.85 − $3.80, or $55.05. The range is given as a single value, $55.05, although in practice it is common to say "the range is $3.80 to $58.85."

For all its simplicity, the worth of the range as a measure of variability should not be underestimated. To know that the range of temperatures to which items stored in a warehouse are subjected is −10 to 120 degrees Fahrenheit can be crucial to deciding what to store there. Similarly, an investment manager choosing between two stocks whose mean closing prices last year were both $50.00 would be most interested to know that one ranged from $15.00 to $90.00 while the other varied between $41.00 and $58.00.

The Standard Deviation

The measure of variability with the greatest statistical impact is the *standard deviation*. The application of the standard deviation in inferential statistics is discussed in later chapters, but its computation is more properly described here, with other measures of variability and central tendency.

For a set of data elements of the type illustrated earlier in this chapter, the standard deviation can be computed by

$$\sigma = \sqrt{\frac{\Sigma(X - \mu)^2}{N}} \tag{3-5}$$

where σ is the standard deviation of a population and X, μ, and N are as previously defined.[5]

The computation of the standard deviation, if it is to be done manually, is made easier by an orderly structuring of data and arithmetic operations, as shown in Table 3-12 for the 24 credit-card sales. The method of computing the standard deviation illustrated in Table 3-12 requires two runs through the data set, the first to compute the mean and the second to

[5]Some other types of data, notably certain probability distributions, use other formulas for the standard deviation.

TABLE 3-12   Computation of the standard deviation of 24 credit-card sales for 1 day (standard method)

Number	X	μ	$X - \mu$	$(X - \mu)^2$
1	3.80	23.55	−19.75	390.0625
2	6.38	23.55	−17.17	294.8089
3	8.21	23.55	−15.34	235.3156
4	10.25	23.55	−13.30	176.8900
5	12.90	23.55	−10.65	113.4225
6	13.29	23.55	−10.26	105.2676
7	14.73	23.55	− 8.82	77.7924
8	14.76	23.55	−8.79	77.2641
9	15.51	23.55	−8.04	64.6416
10	18.33	23.55	−5.22	27.2484
11	19.57	23.55	−3.98	15.8404
12	20.53	23.55	− 3.02	9.1204
13	20.95	23.55	−2.60	6.7600
14	21.70	23.55	−1.85	3.4225
15	23.18	23.55	−.37	.1369
16	24.02	23.55	.47	.2209
17	25.06	23.55	1.51	2.2801
18	31.15	23.55	7.60	57.7600
19	33.40	23.55	9.85	97.0225
20	35.62	23.55	12.07	145.6849
21	38.22	23.55	14.67	215.2089
22	45.35	23.55	21.80	475.2400
23	49.43	23.55	25.88	669.7744
24	58.85	23.55	35.30	1246.0900

$$\Sigma(X - \mu)^2 = 4507.2755$$

$$\sigma = \sqrt{\frac{\Sigma (X - \mu)^2}{N}} = \sqrt{\frac{4507.2755}{24}} = \sqrt{187.8031} = 13.70$$

subtract the mean from every element in the set and to square the difference. Equation (3-5) can be rearranged into a form that does not use the mean and thus permits computation of the standard deviation with a single run. The reworked formula, often called the *shortcut formula,* is

$$\sigma = \sqrt{\frac{\Sigma X^2 - (\Sigma X)^2/N}{N}} \tag{3-6}$$

Many electronic calculators are designed to accumulate the sums of X and X^2 as each value of X is entered. Some of these calculators will then give the standard deviation directly with a keystroke. If you do not have such a calculator but wish to use the shortcut formula manually, a format similar to the one in Table 3-13 may be helpful.

It is no coincidence that equation (3-5) (Table 3-12) and equation (3-6)

TABLE 3-13   Computation of the standard deviation of 24 credit-card sales for 1 day (shortcut method)

Number	X	X^2	Number	X	X^2
1	3.80	14.4400	13	20.95	438.9025
2	6.38	40.7044	14	21.70	470.8900
3	8.21	67.4041	15	23.18	537.3124
4	10.25	105.0625	16	24.02	576.9604
5	12.90	166.4100	17	25.06	628.0036
5	13.29	176.6241	18	31.15	970.3225
7	14.73	216.9729	19	33.40	1115.5600
8	14.76	217.8576	20	35.62	1,268.7844
9	15.51	240.5601	21	38.22	1,460.7684
10	18.33	335.9889	22	45.35	2,056.6225
11	19.57	382.9849	23	49.43	2,443.3249
12	20.53	421.4809	24	58.85	3,463.3225

$$\Sigma X = 565.19 \qquad \Sigma X^2 = 17,817.2645$$

$$\sigma = \sqrt{\frac{\Sigma X^2 - (\Sigma X)^2/N}{N}} = \sqrt{\frac{17,817.2645 - (565.19)^2/24}{24}}$$

$$= \sqrt{\frac{17,817.2645 - 13,309.9890}{24}} = \sqrt{\frac{4507.2755}{24}}$$

$$= \sqrt{187.8031} = 13.70$$

(Table 3-13) give identical values for the standard deviation. The two equations are only different forms of the same mathematical relationships and will always yield the same results within the limits of accuracy to which elements in the equations are expressed.

Although the practical applications of the standard deviation are limited to more sophisticated statistical methods, it is interesting to note that there is a relationship between the standard deviation and its more simple companion measure, the range. When data are normally distributed (a concept explained in the next chapter), the range is approximately six standard deviations. This relationship is sometimes used to make a crude estimate of the standard deviation when a more formal computation is not possible.

Measures of Variability for Grouped Data

Often secondary sources of data will be in summary form, grouped in a frequency distribution, and will not include measures of variability. Just as measures of central tendency can be estimated from such data, so can measures of variability.

The Estimated Range   Estimation of the range from grouped data is based upon an assumption that the boundaries of the classes represent actual elements in the data set. If this is so, the range will be the difference between the upper boundary of the highest class and the lower boundary of the lowest class. This estimate will never be less than the actual range, but it quite frequently will be greater. For the 24 credit-card sales, the range estimated from the frequency distribution in Table 3-4 is $59.99 − $0.00 = $59.99, a larger value than the true range of $58.85 − $3.80 = $55.05.

The Estimated Standard Deviation   The standard deviation can also be estimated from a frequency distribution, but care must be exercised in the adaptation of formulas for the standard deviation. Briefly, a problem exists when one of the expressions in the formula for the estimate is itself an estimate. This would occur if we attempted to measure and square deviations from the mean, because the true mean cannot be computed from a frequency distribution. The correction required when employing an estimate is discussed in Chapter 5.

To estimate the standard deviation from grouped data, then, we shall adapt the formula that does not involve the mean. The modification consists of substituting the class mark CM for X and multiplying it and its square by the class frequency f:

For individual terms: $\sigma = \sqrt{\dfrac{\Sigma X^2 - (\Sigma X)^2/N}{N}}$

For grouped data: $\hat{\sigma} = \sqrt{\dfrac{\Sigma f(CM)^2 - [\Sigma f(CM)]^2/N}{N}}$ (3-7)

where $\hat{\sigma}$ (read "sigma hat") is an *estimate* of σ.

By adding two more columns to Table 3-10, a column for CM^2 and one for $f(CM)^2$, we have all the data required to compute the estimated standard deviation. Such a format, again illustrating the 24 credit-card sales, is found in Table 3-14. The estimate of the standard deviation in this example is smaller than the true standard deviation computed with individual terms in Tables 3-12 and 3-13. While some difference between the true and estimated standard deviation can be expected, larger populations will generally result in smaller differences.

OTHER MEASURES

The median is a measure that divides a data set into two equal parts. There are other measures that divide data sets into more than two equal parts.

TABLE 3-14 *Computation of the estimated standard deviation of 24 credit-card sales for 1 day*

Class	f	CM	CM^2	f(CM)	f(CM)^2
$0.00–$9.99	3	5.00	25.00	15.00	75.00
10.00–19.99	8	15.00	225.00	120.00	1,800.00
20.00–29.99	6	25.00	625.00	150.00	3,750.00
30.00–39.99	4	35.00	1225.00	140.00	4,900.00
40.00–49.99	2	45.00	2025.00	90.00	4,050.00
50.00–59.99	1	55.00	3025.00	55.00	3,025.00
	$N = 24$			$\Sigma f(\text{CM}) = 570.00$	$\Sigma f(\text{CM})^2 = 17{,}600.00$

$$\hat{\sigma} = \sqrt{\frac{\Sigma f(\text{CM}) - [\Sigma f(\text{CM})]^2/N}{N}} = \sqrt{\frac{17{,}600.00 - (570.00)^2/24}{24}}$$

$$= \sqrt{\frac{17{,}600.00 - 13{,}537.50}{24}} = \sqrt{\frac{4062.50}{24}}$$

$$= \sqrt{169.2708} = 13.01$$

Quartiles

Quartiles, as the name implies, divide a data set into *four* equal parts. The first quartile Q_1 separates the first and second parts; the second quartile Q_2 separates the second and third parts; and the third quartile Q_3 separates the third and fourth parts. The second quartile is, of course, identical to the median and can be found at the $[(N + 1)/2]$th position in an array. The first quartile can be thought of as the median of the lower half of an array and is at the $[(N + 1)/4]$th position. The third quartile has a similar position in the upper half of an array, the $[3(N + 1)/4]$th position.

Quartiles can also be computed from a frequency distribution with slight modifications to the formula for the computational median:

$$Q_1 = L_1 + \text{CI}\left(\frac{N/4 - a_1}{b_1}\right) \tag{3-8}$$

$$Q_2 = L_2 + \text{CI}\left(\frac{2N/4 - a_2}{b_2}\right) \tag{3-9}$$

$$Q_3 = L_3 + \text{CI}\left(\frac{3N/4 - a_3}{b_3}\right) \tag{3-10}$$

where L_1, L_2, L_3 are the lower limits of the classes containing Q_1, Q_2, and Q_3, CI is the class interval, and a_1, a_2, a_3 and b_1, b_2, b_3 are analogous to a and b in the formula for the computational median.

The computation of quartiles is left to interested readers, who can check their work against the data in Table 3-15; it shows the actual and

TABLE 3-15   Quartiles for 24 credit-card sales for 1 day

Quartile	Position*	Value from array	Value computed from frequency distribution
Q_1	6.5	14.01	13.75
Q_2	12.5	20.74	21.67
Q_3	18.5	32.28	32.50

*All fractional positions are rounded to .5; thus

$$\frac{24 + 1}{4} = 6.25 \approx 6.5 \quad \text{and} \quad \frac{3(24 + 1)}{4} = 18.75 \approx 18.5$$

computed (from a frequency distribution) quartiles for the 24 credit-card sales.

Deciles and Percentiles

Deciles divide a data set into 10 equal parts, and *percentiles* divide a set into 100 equal parts. Again, there is in each case one less measure than the number of parts, so that deciles will number D_1 to D_9 and percentiles P_1 to P_{99}. Deciles and percentiles are appropriate for much larger populations than we have been using in this chapter, but their interpretations are quite similar to the other equal-part measures. For example, the fifth decile D_5 and the fiftieth percentile P_{50} are both equal to the median. The sixth decile D_6 marks the point between the lower six-tenths and upper four-tenths of the items in a data set, the 83d percentile P_{83} separates the lower 83 percent from the upper 17 percent, and so on.

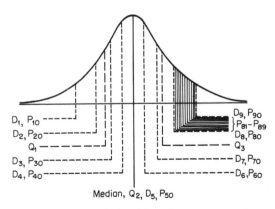

Figure 3-8 Relationships between median, quartiles, deciles, and percentiles.

Percentiles are particularly convenient in comparing a single data element to the population or data set. It is, for example, much more informative to say that a family income is at the 74th percentile than to say that it is $10,000 above the median. Percentiles are also commonly used in describing performance on standardized tests, such as the Scholastic Aptitude Test (SAT) and the Graduate Record Examination (GRE), where a table for converting a standardized test score into a percentile is furnished along with the test results.

We shall not discuss the methods used to compute deciles and percentiles (they are logical extensions of the methods used to compute the median and quartiles) but offer Figure 3-8 to illustrate the relationships between these other measures.

A Closing Note

Most of the material in this chapter is presented for two reasons: it is essential for the understanding and application of more sophisticated statistical methods, and it constitutes part of the basic working knowledge of statistics expected of contemporary managers. However, standing alone and without the opportunity for practice exercises typical of statistics textbooks, some topics may seem irrelevant at this point. If so, don't despair! Everything will eventually fall into place.

CHAPTER 4

Probability and Probability Distributions

In Chapter 1 three decision-making environments were postulated: certainty, uncertainty, and risk. Risk, it will be recalled, is a condition under which probabilities can be assigned to possible outcomes or events. The extent to which decisions are made under risk varies for different businesses, industries, and managerial levels; but even if such decisions are rare, they are often important. For example, the entire insurance industry relies on probabilities drawn from actuarial data; the misuse of these probabilities in a decision by a firm to set new premium rates could be ruinous to that company. In other cases, the decision facing a manager may concern probabilities directly: how to determine probability, how to apply one, or how to interpret the results of a probabilistic method. There are very few cut-and-dried situations when one deals with probabilities; instead, every situation must be analyzed for indications of the relevant probabilistic concepts. This chapter will cover the concepts most frequently required to make intelligent use of probabilities in decision making.

PROBABILITY

Probability is merely a way of expressing the likelihood of an event. Qualitatively, probability can be modified with an adjective expressing the chance of occurrence, such as "remote," "poor," "weak," "good," or "strong." In quantitative terms, probability is expressed as a numerical value, either a fraction or decimal, between 0 and 1. A probability of 0 implies that an event cannot occur and a probability of 1 means that it is

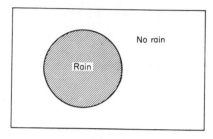

Figure 4-1   Venn diagram of a simple event.

certain to occur. There is also an excellent method, borrowed from set theory, for depicting probability graphically using a *Venn diagram.*[1]

We have become accustomed to hearing the likelihood of rain expressed as a probability, sometimes qualitatively, as in "a slight probability of rain," but more often quantitatively, as in "a 20 percent probability of rain." This probability is illustrated by the Venn diagram in Figure 4-1. Venn diagrams conventionally use a rectangle to represent all events (100 percent) and a circle to represent the probability of some specific event. Although the area of the circle in Figure 4-1 is approximately 20 percent of the area of the rectangle, Venn diagrams are not usually drawn to scale.

Events

The term *event* is used rather loosely in discussing probability. We usually think of an event as an occurrence of some sort, a historical event, a physical phenomenon, a great party, and so forth. In statistics, we broaden that usage to include specified outcomes to probabilistic experiments, such as the suit of a card drawn from a deck, the amount of liquid dispensed by a bottle-filling machine, the fit of bolt threads, and other characteristics that ordinarily would not be described as events.

For the purpose of determining probabilities, it is helpful to classify events as *simple* or *compound.* Briefly, a simple event cannot be subdivided further into component events and a compound event always has two or more component simple events. The event "spade" is simple by this definition, but "ace of spades" could be considered compound because the card must be both an ace (one event) and a spade (a second event). We qualify this example of a compound event because it can also be simple, as can any compound event, if the experimenter chooses to define it so.

[1]Set theory provides a clear, precise, and unambiguous framework for the explanation of probability. It is also guaranteed to drive the casual practitioner up a wall, as anyone with a son or daughter in fifth grade can attest. We therefore avoid the rigorous mathematics of set theory but retain a few features, such as Venn diagrams, that we have found particularly helpful as pedagogical aids.

We shall illustrate this in the discussion of the probabilities of compound events.

The Probability of a Simple Event

All statistical probabilities are expressed as fractions or decimals, but there are significant differences in the origins of those values. For simple events, probabilities may be derived *logically,* observed *empirically,* or assigned *subjectively.* All three methods have important implications for managers.

Logical Probabilities   Some processes are so predictable and so rigidly defined that the probabilities of their outcomes can be deduced logically. Unfortunately, very few business processes fall into this category, but the derivation of logical probabilities is worth examining because the predictability of simple processes sometimes gives us clues to the correct treatment of the unpredictable, complex situations that managers often face.

Throwing dice is one of man's oldest known probabilistic endeavors. We know that the roll of a pair of contemporary dice gives a number from 2 to 12, but what is the probability of, say, a 5? Although there are eleven different outcomes or events and only one of them is a 5, we cannot say that the probability of a 5 is 1/11, because we did not consider the different *ways* events can be generated. Table 4-1 shows a matrix of the possible combinations of two dice. Of the 36 possible combinations, 4 result in the event "5." The probability of a 5 on one roll of two fair dice is therefore 4/36, or approximately .11.

The dice example suggests a general rule for the logical probability of an event:

The logical probability of an event is the ratio of the number of ways that event can occur to the total number of ways all outcomes to an experiment can be generated.

TABLE 4-1 *The four ways in which a 5 can occur from rolling two dice*

Value of second die	Value of first die					
	1	2	3	4	5	6
1	2	3	4	5	6	7
2	3	4	5	6	7	8
3	4	5	6	7	8	9
4	5	6	7	8	9	10
5	6	7	8	9	10	11
6	7	8	9	10	11	12

Or, in more concise notation,

$$P(\text{event}) = \frac{\text{number of ways event can occur}}{\text{number of ways all outcomes can occur}}$$

Empirical Probabilities Most situations confronting managers do not follow such apparent probabilistic patterns. The situation may, indeed, be probabilistic; it just is not apparent what the probabilities are. In such cases, it is appropriate to refer to observed, or empirical, data.

Suppose that a recent production run of 10,000 integrated circuits of a certain type yielded 25 that were flawed. On the basis of these empirical data, we would say that the probability of a flawed integrated circuit is 25/10,000 = .0025. Probability in this case is both the likelihood that 1 random unit will be flawed and the expected percentage of flawed units in a batch. In an order of 2000 integrated circuits from this firm we could expect .0025(2000) = 5 flawed units.

Although the concept of probability is the same for logical and empirical processes, empirical probabilities can be understood better if the rule for logical probabilities is modified slightly:

The empirical probability of an event is the ratio of the observed frequency of that event to the total number of observations.

Or, in the concise form,

$$P(\text{event}) = \frac{\text{observed frequency}}{\text{total frequency}}$$

Subjective Probabilities When a process is not predictable and observed data are unavailable, a manager frequently must assign probabilities subjectively. This assignment may be based on qualitative factors, experience in similar situations, and even intuition. For example, weather does not follow a known, probabilistic pattern, nor can a meteorologist run a few thousand trials of a given set of conditions just to see how often they produce rain. Instead, he considers the winds, pressure areas, adjacent weather systems, and his theoretical knowledge of cause-and-effect relationships in meteorology, adds (one suspects) not a little intuition, and announces his estimate of the probability of rain.

Managers follow similar processes in determining the probability of demand for a product, the probability of success of a space probe, the probability of recovery from a serious illness, and in thousands of other situations. This is a perfectly acceptable managerial technique. However, whether subjective probabilities should be used in further statistical computations is a subject of some controversy. Statisticians of the *objectivist,* or *classical,* school reject such use while those of the *bayesian* school accept it. We shall not presume to resolve the argument here, except to say that the use of subjective probabilities seems to be more appropriate in business decision making than, say, research chemistry. That being the case and this being a book for managers (but certainly not for research chemists), we cautiously promote the use of subjective probabilities.

The Probability of a Compound Event

There are two broad categories of compound events with which we shall be concerned. In the first, *every* specified simple event must occur to satisfy the compound event. In the second category, the occurrence of *any one* or *any combination* of simple events satisfies the compound event.

The probability of a successful trip in an automobile as opposed to the success of a spacecraft subsystem illustrates the difference between the two types of compound events. In order for an automobile to work properly, every one of a number of key subsystems must work. The engine must run, the transmission must not fail, the steering gear must function, and so on. But in a spacecraft, there are usually redundant subsystems. Considering only the guidance computer, the operation of any one (or any combination) of three identical computers will satisfy the compound event "the computer works."

Probabilities of compound events of the first kind are determined by *Laws of Multiplication* while those of the second kind follow *Laws of Addition.* The laws of probability are aptly named; in each case the name indicates the arithmetic operation involved in the determination of a compound event probability.

The Laws of Multiplication

The probability that two events will both occur is called a *joint probability.* The term relates well to a Venn diagram of a compound event where the joint probability is depicted by the juncture of the circles representing the two simple events. A Venn diagram of the compound event "two spades in two draws from a card deck without replacement" is shown in Figure 4-2.

The probability of the simple event "spade on the first draw" is given by the ratio of spades to the total number of cards, 13/52. The probability of the second simple event is more difficult to determine. Since the com-

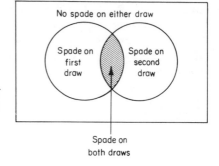

Figure 4-2   Venn diagram of the probability of a compound event (multiplication).

pound event specifies "without replacement," the probability of a spade on the second draw is dependent on the outcome of the first draw. If a spade was drawn on the first attempt, only 12 of the remaining 51 cards are spades and the probability of a spade on the second attempt is 12/51. However, if a spade was not drawn on the first attempt, all 13 spades remain and the probability is 13/51. Probabilities dependent on another event in this manner are called *conditional probabilities*.

The *General Law of Multiplication* states that the joint probability of two events is the product of the probability of the first event and the probability of the second event, *given that the first event has occurred.* Designating a spade on the first draw as S_1 and a spade on the second draw as S_2, and using a vertical line | to represent "given," we determine the joint probability of two spades in two draws:

$$P(S_1) \times PS_2 \mid S_1 \text{ and } S_2) = P(S_1 \times P(S_2 \mid S_2)$$
$$= 13/52 \times 12/51$$
$$= 156/2652 = .0588$$

It is interesting to note that the word "and," usually associated with addition, implies multiplication in probability. In this case, "and" is actually used in a very narrow sense as one of the operative terms in the branch of mathematics called *boolean algebra*, which is commonly used in set theory.

The General Law of Multiplication *always* works for joint probabilities. However, there are some special cases in which a simpler form of the multiplication law applies. If the compound event of drawing two spades specifies "with replacement," the probability of S_2 is the same whether S_1 occurs or not, because the first card is back in the deck. In this case S_1 and S_2 are said to be independent, and $P(S_2) = P(S_2 \mid S_1)$. This gives rise to the *Special Law of Multiplication*, which states that the joint probability of two independent events is equal to the product of the probabilities of the simple events. For the joint probability of two spades in two draws with replacement,

$$P(S_1 \text{ and } S_2) = P(S_1) \times P(S_2)$$
$$= 13/52 \times 13/52$$
$$= 169/2704 = .0625$$

In practice, the Special Law of Multiplication can be used instead of the General Law when the difference between an unconditional and a conditional probability is slight. For example, a machinist picks two bolts at random from a bin of 10,000 bolts, 100 of which are flawed; what is the probability that both selected bolts will be flawed? Since 100 bolts are flawed, the probability of a flawed bolt is 100/10,000, or .01. If the first bolt is flawed, the probability of a second flawed bolt is only 99/9999, or .0099, and the probability of flaws on both bolts is .01 × .0099, or .000099. In practice, we would most likely use the Special Law of Multiplication and assume that the probability of two flawed bolts is .01 × .01, or .0001.

The Laws of Addition

While Laws of Multiplication typically apply when *both* (or all) simple events must occur to satisfy the compound event, Laws of Addition apply when *either* (or any) simple event satisfies the compound event. To continue the playing-card example, let us specify a compound event of "spade on the first or second draw with replacement." Now the key word is "or," also a boolean operator and one which signals addition. The compound event is satisfied by a spade on the first draw, a spade on the second draw, or a spade on both draws, as shown in Figure 4-3.

Although the application of a Law of Addition suggests that the probabilities of S_1 and S_2 be added, it can be seen from Figure 4-4 that mere addition will overstate the probability of S_1 or S_2 because the area of joint probability will be counted twice. To compensate for the overlap of the simple events, the *General Law of Addition* states that the probability of either of two events is equal to the sum of the probabilities of each of the

Figure 4-3   Venn diagram of the probability of a compound event (addition).

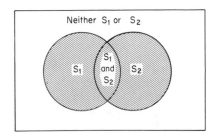

events *minus their joint probability.* Thus, the probability of a spade on either the first or second draw is given by

$$
\begin{aligned}
P(S_1 \text{ or } S_2) &= P(S_1) + P(S_2) - P(S_1 \text{ and } S_2) \\
&= 13/52 + 13/52 - (13/52 \times 13/52) \\
&= \frac{676}{2704} + \frac{676}{2704} - \frac{169}{2704} \\
&= \frac{1183}{2704} = .4375
\end{aligned}
$$

At this point, some readers may be asking: What if the compound event were defined *without* replacement? Obviously the joint probability would be determined from the General instead of the Special Law of Multiplication, but what probability would be used for S_2: 12/51 (indicating that S_1 occurred) or 13/51 (indicating that S_1 did not occur)? The curious fact is that the occurrence of S_1 is not a factor in this case and the probability of S_2 is 13/52, just as if the first card had never been drawn! The proof that the probability of S_2 is 13/52 is beyond the scope of this book, and we shall say only that it is based on something called the *Principle of Insufficient Reason.*

Just as there is a Special Law of Multiplication, so there is a *Special Law of Addition,* but its applicability is determined not by independence but by a new concept, whether or not events are *mutually exclusive.*

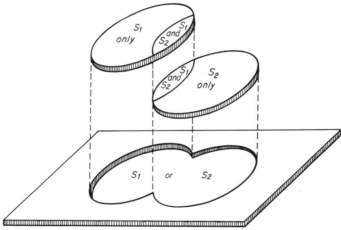

Figure 4-4 The area of overlap between events that are not mutually exclusive.

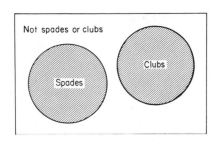

Figure 4-5   Venn diagram of mutually exclusive events.

Events are said to be mutually exclusive when the occurrence of one event absolutely precludes the occurrence of the other(s). For the drawing of one card, the event "spade" precludes the events "heart," "diamond," and "club." A "four" on one roll of a die precludes the occurrence of any other number, and so on. While there is no pictorial representation of independence, mutually exclusive events are clearly distinguished in a Venn diagram by nonoverlapping circles (Figure 4-5).

To complete the demonstration of the laws of probability with playing cards, consider the compound event "spade or club on one draw from a full deck." As we suggested earlier, a compound event can be defined in such a way to make it a simple event. We could have specified "black suit" instead of "spade or club" and determined the probability to be 26/52, since we know there are 26 cards in the two black suits. Most applications in management will not be so obvious, however, and we shall continue with "spade and club" to set an example for more complex situations.

The absence of an overlap area simply means that the joint probability of the two events is 0; they cannot both occur. When we substitute 0 for the joint-probability term in the General Law of Addition, it drops out and we are left with the Special Law:

The probability of either of two mutually exclusive events is equal to the sum of the probabilities of each.

For the probability of either a spade or club on one draw,

$$P(\text{S or C}) = P(\text{S}) + P(\text{C})$$
$$= 13/52 + 13/52$$
$$= 26/52 = .5000$$

The probabilities of compound events consisting of more than two simple events will not be discussed, but the interested reader can derive the computational procedures from a study of an appropriate Venn diagram and an extension of the logic incorporated in the two event examples.

Bayes' Law

The General Law of Multiplication includes a conditional probability of $B \mid A$ as in

$$P(A \text{ and } B) = P(A)P(B \mid A) \tag{4-1a}$$

Whether A or B is the first event is immaterial and the General Law can also be written as

$$P(A \text{ and } B) = P(B)P(A \mid B) \tag{4-1b}$$

With a little algebraic manipulation, these two equations can be combined to form a third, which includes the conditional probabilities of both $B \mid A$ and $A \mid B$:

$$P(A \mid B) = \frac{P(A)P(B \mid A)}{P(B)} \tag{4-1}$$

Equation (4-1), which is one form of *Bayes' law,* gives a unique opportunity of finding a conditional probability when its reverse and the unconditional probabilities are known. For example, let us take the case of a soap manufacturer who periodically repackages powdered soap in different boxes under different brand names. In the past, a repackaged soap has had a .40 probability of becoming what the manufacturer terms a "nationwide success." Before undertaking nationwide distribution, this manufacturer usually tests a new soap in Los Angeles, where the probability of success for new soaps has been .50. According to company records, 80 percent of the soaps that went on to be nationwide successes were successful in Los Angeles. A new soap under the name "Magic" in a blue box with white stars and cresent moons is tested in Los Angeles and is an instant success. How can the manufacturer now revise the estimate of .40 for a nationwide success?

$$
\begin{aligned}
\text{Let } A &= \text{success nationwide} & P(A) &= .40 \\
B &= \text{success in Los Angeles} & P(B) &= .50 \\
B \mid A &= \text{Los Angeles success given} & P(B \mid A) &= .80 \\
&\quad \text{nationwide success} \\
A \mid B &= \text{nationwide success given} & P(A \mid B) &= ? \\
&\quad \text{Los Angeles success}
\end{aligned}
$$

According to Bayes' law,

$$P(A \mid B) = \frac{P(A)P(B \mid A)}{P(B)}$$

and the probability of a nationwide success given a success in Los Angeles is

$$\frac{(.40)(.80)}{.50} = .64$$

The revised probability of nationwide success should now be used in the preparation of pro forma profit-and-loss statements, production schedules, advertising budgets, and other similar managerial tools.

The principle of revising probabilities based on test results is well known in marketing and has applications in other functional areas. The revision of subjective probabilities on the basis of empirical (sample) data is a particularly significant aspect of bayesian decision analysis.

PROBABILITY DISTRIBUTIONS

So far, probabilities have been expressed only in terms of one specified event, other possible events being considered only to the extent to which they determine the denominator in the probability ratio. The other, unspecified events have probabilities too. When these additional probabilities are determined, a *probability distribution* of all possible events in an experiment can be formed. For the simple experiment of casting one die, the probability distribution consists of equal probabilities for the six possible events, as shown in Table 4-2.

TABLE 4-2   The probability distribution for one roll of a die

Event	P(event)
1	$\frac{1}{6}$
2	$\frac{1}{6}$
3	$\frac{1}{6}$
4	$\frac{1}{6}$
5	$\frac{1}{6}$
6	$\frac{1}{6}$

Probability distributions are similar to relative frequency distributions (Chapter 3) in that both relative frequency and probability total 1.000 for a given distribution. In fact, the relative frequency distribution of a very large number of rolls of one die would be indistinguishable from the probability distribution in Table 4-2. In this case, the relative frequency distribution would show empirical probabilities while Table 4-2 shows logical probabilities.

Distributions of probabilities are helpful to managers because they give some idea of what happens when the specified event does not occur. An automobile manufacturer's initial appraisal of the market for a new model may be "a 40 percent chance of 300,000 sales," a subjective probability. But what of the remaining 60 percent? Upon consideration of additional economic factors, competitors, plans, and market surveys, the probabilities of other sales levels could be estimated to complete the probability distribution of demand shown in Table 4-3. This distribution gives a much more complete picture of demand than the probability of one quantity does.

TABLE 4-3   The probability distribution of demand for a new automobile model

Demand	P (demand)
100,000	.10
200,000	.25
300,000	.40
400,000	.15
500,000	.10

The demand data in Table 4-3 can be combined with other information to form a probability distribution of profit. If each automobile sold contributes $500 and fixed costs are $125 million, profit will be distributed as shown in Table 4-4. Faced with a .35 probability of a net loss on this model, the manufacturer may well decide not to market it, although the .40 probability of 300,000 sales ($25 million profit) may have seemed attractive initially.

TABLE 4-4   The probability distribution of profit on a new automobile model

Millions of dollars				
Contribution*	− Fixed cost	= Profit		P(profit)
50	125	−75		.10
100	125	−25		.25
150	125	25		.40
200	125	75		.15
250	125	125		.10

*Contribution = demand × $500.

There still may be a need for a single demand estimate. The computational processes for determining capital budgets, sticker price, and the quantities in contracts with suppliers all require a single value for the expected demand. In most cases, the best single measure of a probability distribution is the *expected value,* which corresponds to the mean of a frequency distribution and is computed in much the same way. The expected value of demand for the automobile example is shown in Table 4-5 and suggests a general formula for the expected value of any distribution of a variable X:

$$\text{Expected value of } X = \text{EV}(X) = \Sigma XP(X) \tag{4-2}$$

The expected demand of 290,000 is not that much different from the original estimate of 300,000; within the tolerances with which one can make such estimates they are, for all practical purposes, the same. Why then go to the trouble of constructing a probability distribution? The answer is that it is not a matter of a single estimate versus an expected value but a question of using the best, most complete data wherever possible. In this case, the probability distribution gives a better profit-and-loss picture, but it is not convenient to use in some other computations.

The probability of demand for automobiles is unique to a given set of conditions: a certain manufacturer, the characteristics of the automobile itself, the current state of the economy, the cost and availability of gasoline, and so forth. In such cases it is appropriate to form a distribution of subjective probabilities. However, other events have been found to follow definite probabilistic patterns and can be described by one of several theoretical probability distributions. We shall discuss three distributions that have particular significance to managerial decision making.

TABLE 4-5   The expected value of demand for a new automobile model

Demand	P(demand)	Demand × P(demand)
100,000	.10	10,000
200,000	.25	50,000
300,000	.40	120,000
400,000	.15	60,000
500,000	.10	50,000
		E.V.(demand) = 290,000

The Binomial Probability Distribution

The binomial probability distribution is one of two discrete distributions that describe many common business processes and physical phenomena. For the binomial distribution to apply, four conditions must exist:

- The process must have, or be capable of being defined as having, two and only two mutually exclusive, collectively exhaustive events.

- The probability of an event must be the same for every trial and may not vary with time or the number of trials.

- Trials must be independent; that is, the outcome of one trial cannot be influenced by the outcome of a previous trial.[2]

- There must be a discrete number of trials.

Once again, we shall use a die to illustrate probability concepts, but first let us confirm that casting a die meets the prerequisites of the binomial probability distribution.

A die has six faces, and one roll of a die can result in one of six different events. But the outcome of that roll can be *defined* to meet the two-event condition. We shall define the specified event as "four" and the other event as "not four." These events are obviously mutually exclusive, and, equally important, they are collectively exhaustive; that is, "four" and "not four" cover all possible outcomes of rolling one die.

The probability of rolling a 4 will not vary, assuming the die is fair. Even after thousands of rolls, we would expect wear on the edges and faces to be equally distributed and the probability of a four to be unchanged at $1/6$.

A die has no memory. In the unusual circumstance of, say, five straight 4s, the probability of a 4 on the next roll is still $1/6$. The trials, or rolls, are independent.

Finally, there is a distinct notion of individual trials in the casting of dice. Each roll is a separate trial, clearly distinguished from previous and future rolls.

Parameters of the Binomial Probability Distribution   It is customary to refer to *the* binomial distribution as if there were only one. Actually, there is an infinite number of binomial distributions, one for every possible combination of the number of trials n and the probability p that the specified event will occur on any given trial. These two factors, which

---

[2]This condition is actually implied in the second condition since, by definition, independence exists when $P(B|A$ occurs$) = P(B|A$ does not occur$)$. If the probability of an event does not change, it must be independent of previous events.

define a specific binomial distribution, are called the *parameters* of the distribution.

The number of times the specified event occurs out of n trials is a *random variable* and is designated x. Table 4-6 shows the binomial probability distribution of x given $n = 10$ and $p = \frac{1}{6}$, which is the distribution that describes the probability of x 4s in ten rolls of a fair die.

TABLE 4-6   *The binomial probability distribution for* $n = 10$, $p = \frac{1}{6}$

x	$P(x)$	x	$P(x)$
0	.1615	6	.0022
1	.3230	7	.0002
2	.2907	8	.00002
3	.1550	9	.000001
4	.0543	10	.00000002
5	.0130		

Tables of Binomial Probabilities   Appendix D gives the binomial probabilities of x for various combinations of n and p. Although binomial probabilities can be computed by hand (or with a scientific calculator), using tables is faster and easier for most people. The following examples demonstrate the application of tables to different expressions of binomial probabilities:

$$P(x = 3 | n = 8, p = .15) = .0839$$

Table for $n = 8$, column for $p = .15$, row for $x = 3$.

$$P(x = 2 | n = 10, p = .30) = .2335$$

Table for $n = 10$, column for $p = .30$, row for $x = 2$.

$$P(x \leq 2 | n = 10, p = .30) = .3828$$

Since ≤ 2 includes 0, 1, and 2, $P(x \leq 2)$ is found from the sum of

$P(x = 0) + P(x = 1) + P(x = 2)$, or $.0282 + .1211 + .2335 = .3828$.

$$P(x > 5 | n = 7, p = .45) = .0357$$

Since >5 includes 6 and 7, $P(x > 5)$ is found from the sum of

$P(x = 6) + P(x = 7)$, or $.0320 + .0037 = .0357$.

$$P(x \leq 7 | n = 9, p = .50) = .9804$$

This probability could be found by the sum of $P(x = 0) + P(x = 1) + \cdots + P(x = 7)$. However, since the sum of all probabilities in the distribution is 1.0000, it is easier to subtract the unwanted terms from 1.0000: $P(x \leqslant 7) = 1.0000 - [P(x = 8) + P(x = 9)]$ or $1.0000 - (.0176 + .0020) = .9804$.

$$P(x = 5|n = 8, p = .65) = .0808$$

Probabilities for values of p greater than .50 cannot be read from the tables in Appendix D, but it is possible to convert a binomial distribution for which $p > .50$ into one for which $p < .5$ by the formula

$$P(x|n, p) = P(n - x|n, 1 - p)$$

Thus, $P(x = 5|n = 8, p = .65)$ is the same as $P(x = 3|n = 8, p = .35)$, and the probability can be read directly from the tables as .0808. The similarity of these two distributions ($n = 8, p = .65$ and $n = 8, p = .35$) that facilitates the conversion is illustrated in Figure 4-6.

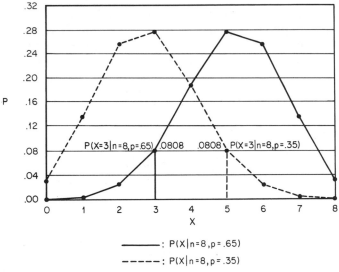

Figure 4-6   Binomial probability distributions of $n = 8, p = .65$ and $n = 8, p = .35$.

Applications of the Binomial Probability Distribution   There are many "two-event" situations in business that can be described by the binomial distribution. A customer either makes or does not make a purchase, a quality control inspector either passes or does not pass a piece of glassware, an egg is either grade A or it is not, and so on. If the probability of one event and independence can be established, the probability of x events out of n trials can be determined with the binomial distribution.

It should be noted at this point that not all tables of binomial probabilities are constructed exactly like those in Appendix D. Some use π for p, k for x, or make other changes in notation. Another common variation is to construct cumulative (decumulative) tables that give the probability that x is less than (greater than) some number. Before using a set of tables, one should always take careful note of the symbols used, the type of construction used (cumulative, decumulative, or individual terms), the range of p (beyond .50 or not), and any instructions for the use of the tables that may be found in the accompanying text.

The Mean and Standard Deviation of the Binomial Probability Distribution   In addition to the variation in table construction, there is also a great deal of variation in table length; that is, the values of n included in the table. Regardless of the length of the table (Appendix D shows values of n from 1 to 20, 25, and 30), there will be times when the tables at hand do not cover the value of n in the problem to be solved. In such cases one can usually approximate the binomial distribution with another distribution.

The link between the binomial distribution and other distributions is its mean and standard deviation. These measures have the same relationship to a binomial probability distribution as they do to a frequency distribution, but because of the special characteristics of the binomial distribution they can be calculated much more easily.[3] Although the mean and standard deviation of the binomial distribution are not needed at this time, they are included here, with other information on the binomial distribution, for later use:

$$\text{Binomial mean} = \mu_B = np \qquad (4\text{-}3)$$
$$\text{Binomial standard deviation} = \sigma_B = \sqrt{np(1 - p)} \qquad (4\text{-}4)$$

The Poisson Probability Distribution

A second discrete distribution, and one closely related to the binomial, is the Poisson probability distribution. Both distributions describe what are known as *Bernoulli* processes, but there is no concept of distinct trials in

---

[3]They can also be calculated as they are for a frequency distribution, using the values of x as class marks and the probabilities of x as frequencies. This is sometimes an interesting classroom exercise but has little value to practicing managers.

the Poisson. Thus, the conditions under which the Poisson distribution applies are quite similar to those for the binomial distribution:

- The process must be a "two-event," or Bernoulli, process.

- There must be some mean number of specified events per measure of time or space that does not change for the duration of the process.

- The process must be continuous, without individual trials.

To illustrate the Poisson distribution, suppose that a weaving process produces cloth with an occasional flaw. The cloth is either flawed or not flawed (a Bernoulli process) and the number of flaws can be counted easily. The number of "nonflaws" cannot be counted, however, nor can the number of opportunities (trials) for a flaw to exist. Therefore, there is no probability of a flaw as such, only a mean number of flaws per unit of area, say, three flaws per square yard.

Parameter of the Poisson Probability Distribution Since there are no distinct trials in a Poisson process, there can be no parameters n or p for the Poisson distribution. Instead, there is a single parameter λ, the mean number of occurrences per measure of time, distance, area, or volume. Given a *mean* number of occurrences, the *actual* number of occurrences is a random variable that we designate X. The distribution of Poisson probabilities of X for $\lambda = 3$ is shown in Table 4-7.

*TABLE 4-7   The Poisson
probability distribution for $\lambda = 3$*

X	$P(X|\lambda = 3)$	X	$P(X|\lambda = 3)$
0	.0498	7	.0216
1	.1494	8	.0081
2	.2240	9	.0027
3	.2240	10	.0008
4	.1680	11	.0002
5	.1008	12	.0001
6	.0504		

For the cloth-weaving process, Table 4-7 shows that the probability of no flaws in a square yard is .0498, the probability of only one flaw is .1494, and so on.

The binomial distribution has a definite upper limit for the random variable: x obviously cannot exceed n. But the Poisson has no n and, in theory, no upper limit to X. In practice, one usually ignores, as we have in Table 4-7, values of X with probabilities less than .0001.

Tables of Poisson Probabilities   Appendix E contains tables of the Poisson probabilities of X for various values of λ. Again, while there are methods of computing Poisson probabilities, tables are convenient and easy to use, as shown in the following examples.

$$P(X = 4|\lambda = 2.5) = .1336$$

Column for $\lambda = 2.5$, row for $X = 4$.

$$P(X \leq 3|\lambda = 4.2) = .3955$$

Since $X \leq 3$ includes $X = 0$, $X = 1$, $X = 2$, and $X = 3$, the sum of $P(X = 0) + P(X = 1) + P(X = 2) + P(X = 3)$ will give $P(X \leq 3)$: .0150 + .0630 + .1323 + .1852 = .3955.

$$P(X > 17 = 8.5) = .0030$$

In this case, the probabilities of values of X larger than 17 (18, 19, 20, 21, and 22) are summed to give .0030.

$$P(X < 13|\lambda = 5.7) = .9947$$

This probability could be found by summing the individual terms from 0 to 12, but it is easier to subtract from 1.0000 the probabilities of $X = 13$, 14, 15, 16, and 17:

$$1.0000 - (.0032 + .0013 + .0005 + .0002 + .0001) = .9947$$

Applications of the Poisson Probability Distribution   The example of flaws in a piece of cloth is only one of many common events that follows the Poisson distribution. The arrival of automobiles at a service station, patients at a hospital, or secretaries at a duplicating machine may also be described by the Poisson distribution. In some cases, the assumptions of a Poisson process should be questioned before an important decision is based on a Poisson probability. It may be that the mean number of flaws per square yard *does* change as machine settings get out of tolerance; or hospital arrivals may *not* be independent, since automobile accidents, for example, frequently produce several victims and the arrival of the first greatly increases the probability of a second or third. A good deal of judgment and qualitative analysis must be exercised before this, or any other, quantitative technique is applied.

While tables of binomial probabilities are limited by the values of n, Poisson tables are somewhat limited by values of λ. We therefore caution against the temptation to scale down probabilities, that is, to assume that the probability of, say, 30 arrivals per hour is 3 times as great as the probability of 10 per hour. It is not, and care must be exercised to avoid this trap. A form of scaling is permitted, however, in the determination of λ. For example, when the mean number of flaws per yard is 3, we can expect the mean for 2.5 yards to be $3(2.5) = 7.5$ and the probability of X flaws in 2.5 yards is found in the column for $\lambda = 7.5$.

Tables of Poisson probabilities also vary somewhat in form from text to text, although not quite as much as binomial tables. The same precautions recommended for the use of other binomial tables also apply to the use of other Poisson tables.

The Mean and Standard Deviation of the Poisson Distribution   Like the binomial distribution, the Poisson distribution can be approximated by another distribution when the table entry value (λ for the Poisson distribution) is too large for the table. Again, the mean and standard deviation are required for the approximation. For the Poisson distribution, these measures are

$$\text{Poisson mean} = \mu_P = \lambda \qquad (4\text{-}5)$$
$$\text{Poisson standard deviation} = \sigma_P = \sqrt{\lambda} \qquad (4\text{-}6)$$

The Poisson Approximation to the Binomial Distribution   When n is too large for the binomial tables, a binomial probability can be approximated with the Poisson distribution if p is sufficiently small. Since both a large n and small p are involved, rules for approximating the binomial distribution with the Poisson distribution are usually expressed in terms of the product of these two parameters. The commonly accepted rule is that the approximation is permitted when

$$np \leq 5$$

As an example of a Poisson approximation to the binomial, consider the probability that exactly 98 of 100 flashbulbs will work when the probability that any given bulb will work is .99. We want the binomial probability $P(x = 98 | n = 100, p = .99)$, but the binomial tables do not go to $n = 100$.

At first glance, it also seems that the Poisson approximation is inappropriate, since $np = 99$, which certainly exceeds 5. We note, however, that exactly 98 successes is the same as exactly 2 failures and a probability of success of .99 implies a probability of failure of .01. The probability, therefore, can be expressed as $P(x = 2|n = 100, p = .01)$, and since np is only 1, the Poisson approximation can be used.

To convert the binomial distribution into the Poisson distribution, one need only substitute the binomial mean np for the Poisson mean λ and solve for X instead of x. In this case, $P(x = 2|n = 100, p = .01)$ is the same as $P(X = 2|\lambda = 1)$, which can be found[4] in Appendix D as .1839.

The Normal Distribution

The normal distribution is different from both the binomial and Poisson distributions in that it is a *continuous* rather than discrete distribution. Variables that can assume any fractional or decimal values between whole numbers, such as weights, lengths, heights, distances, times, and other continuous measures, frequently follow the normal distribution.

In a discrete probability distribution, the probability associated with each value of the random variable can be represented by the height of a line on a graph (see Figure 4-6). In a continuous probability distribution, probabilities are represented by an area under a curve. Thus, we can find the probability that the random variable is greater than or equal to some value, less than or equal to some value, or between two values—but not the probability that it is exactly equal to some value.[5]

Parameters of the Normal Distribution   The parameters of the normal distribution are its mean μ and standard deviation σ. These measures are computed from the data set that constitutes the distribution, as described in the preceding chapter, and therefore need no distinguishing subscripts like those used for the mean and standard deviation of the binomial and Poisson distributions.

The relationship between area measurement of probability in the normal distribution and its two parameters is shown in Figure 4-7. The normal distribution is symmetrical about the mean, and the probability that

---

[4]The true binomial probability $P(x = 2|n = 100, p = .01)$ is .1849.

[5]An exact value occupies a point on a one-dimensional line or a line on a two-dimensional graph of the normal curve. Since a line has only length but not area, the probability of an exact value must be 0 in a continuous distribution.

the random variable is equal to or greater than (equal to or less than) the mean is .5000. The first standard deviation on either side of the mean includes 34.13 percent of the elements in the distribution, and the probability of the random variable's falling *within* one standard deviation is therefore 2 × .3413 = .6826. Similarly, the probability of being within two standard deviations of the mean is 2 × (.3413 + .1359) = .9544 and within three standard deviations it is 2 × (.3413 + .1359 + .0215) = .9974. Only 2 × .0013 = .26 percent of the elements are more than three standard deviations from the mean.

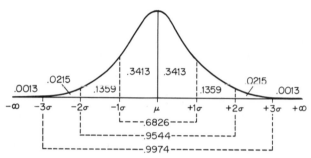

Figure 4-7   Areas under the normal probability curve.

Table of Normal Probabilities   The relationship between probability and distance from the mean measured in standard deviations enables us to get by with only one table of normal probabilities. That table is for the distribution, called the *standard normal distribution,* with a mean of 0 and a standard deviation of 1.

To use the table of standard normal probabilities (Appendix E), one merely converts the random variable into a measure of standard deviations from the mean, a new variable identified as z, with the formula

$$z = \frac{X - \mu}{\sigma} \tag{4-7}$$

For example, let us say that a poultry processor has found that the weights of dressed turkeys are normally distributed with a mean of 12 pounds and a standard deviation of 2.5 pounds. What is the probability that a turkey selected at random will weigh between 12 and 15 pounds?

It is always helpful to sketch the desired area under the normal curve when solving a problem like this one. Figure 4-8*a* shows that the area representing the probability of a weight between 12 and 15 pounds is bounded by the mean and a line at $X = 15$. Since the table of standard

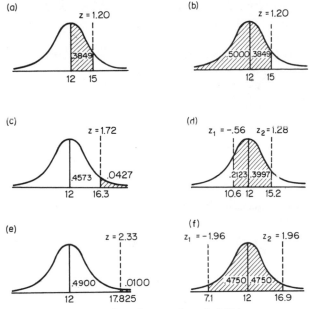

Figure 4-8   Examples of normal probabilities.

normal probabilities is based on distances from the mean, we need only find the z value for $X = 15$ and look up the corresponding probability:

$$z = \frac{15 - 12}{2.5} = 1.20 \qquad P(12 \leq X \leq 15 | \mu = 12, \sigma = 2.5) = .3849$$

The number .3849 is found at the intersection of the 1.2 row and the .00 column in Appendix E.

When the area desired is not bounded by the mean, some additional computations are necessary to obtain the correct probability. The following examples, based on the distribution of turkey weights, illustrate the most common situations.

What is the probability of a weight less than 15 pounds?[6]

$$z = \frac{15 - 12}{2.5} = 1.20$$
$$P(X < 15 | \mu = 12, \sigma = 2.5) = .5000 + .3849 = .8849$$

---

[6]Since the probability of a weight of *exactly* 15 pounds is zero, we do not distinguish between the terms "less than" and "equal to or less than" when working with a continuous distribution.

Figure 4-8*b* shows that the desired area includes all that is less than the mean (.5000) plus that between the mean and $X = 15$.

What is the probability of a weight greater than 16.3 pounds?

$$z = \frac{16.3 - 12}{2.5} = 1.72$$
$$P(X > 16.3 | \mu = 12, \sigma = 2.5) = .5000 - .4573 = .0427$$

Since the table value for a z of 1.72 (found at the intersection of the 1.70 row and .02 column) is the area between the mean and $X = 16.3$, it must be subtracted from the area to the right of the mean (.5000) to give the shaded area in Figure 4-8*c*.

What is the probability of a weight between 10.6 and 15.2 pounds?

$$z_1 = \frac{10.6 - 12}{2.5} = -.56 \qquad z_2 = \frac{15.2 - 12}{2.5} = 1.28$$
$$P(10.6 \le X \le 15.2 | \mu = 12, \sigma = 2.5) = .2123 + .3997 = .6120$$

This problem, illustrated in Figure 4-8*d*, requires two steps. First, the area between the mean and $X = 10.6$ must be found and then the area between the mean and $X = 15.2$. Since the normal distribution is symmetrical, the area between the mean and $-.56$, the z value for $X = 10.6$, is the same as the area between the mean and $+.56$, that is, .2123. The area between the mean and $X = 15.2$ is the area associated with a z of 1.28 or .3997. This problem *must* be worked in two steps, converting the difference between 10.6 and 15.2 into a single z value will not give a correct answer.

What weight will be exceeded by only 1 percent of the turkeys?

Figure 4-8*e* shows that the desired weight is one that separates the top 1 percent from the lower 99 percent. The lower 99 percent consists of the 50 percent less than the mean and the 49 percent between the mean and X. From Appendix E, the table value closest to .4900 is .4901, which is found at the intersection of the 2.3 row and the .03 column. The z value, then, is 2.33, and X can be computed:

$$z = \frac{X - \mu}{\sigma} \qquad 2.33 = \frac{X - 12}{2.5}$$
$$X - 12 = (2.33)(2.5) \qquad X = 12 + 5.825 = 17.825$$

We therefore expect only 1 percent of the turkeys to weigh more than 17.825 pounds.

What symmetrical bracket about the mean will include 95 percent of the turkeys?

Figure 4-8*f* shows that the bracket includes 47.5 percent on each side of the mean. From Appendix E, the *z* value for .4750 can be read as 1.96. By symmetry, the .4750 to the left of the mean is associated with a *z* of −1.96. The bracket can now be calculated:

$$z = \frac{X - \mu}{\sigma} \qquad \pm 1.96 = \frac{X - 12}{2.5}$$

$$X - 12 = (\pm 1.96)(2.5) \qquad X = 12 \pm 4.9 = 7.1 \text{ and } 16.9$$

The 95 percent bracket is from 7.1 to 16.9 pounds.

Applications of the Normal Probability Distribution   With a little imagination, the weights of turkeys can easily become diameters of ball bearings, tensile strengths of rivets, miles of wear on tires, monthly sales revenues, or other similar data used in managerial decision making. If these data are normally distributed (and they frequently are), the probability of any single data element falling short of, beyond, or between specified limits can readily be determined.

The normal distribution is particularly applicable to distributions of measurement errors. Indeed, the normal distribution is frequently called the *gaussian* distribution, for Carl Gauss, who found that errors in certain astronomical measurements were normally distributed. While not many managers have occasion, as Gauss did, to measure planetary orbits, it is common to measure machine settings, temperatures, pressures, and other process variables. The knowledge that errors in these measurements are normally distributed can be most helpful in production control methods.

In the turkey problem, the table of standard normal probabilities was used to find weights that separated or bracketed certain percentages of the turkeys in the distribution. Problems of this type tend to use a few common *z* values associated with frequently used percentages. These *z* values, shown in Table 4-8, are also used often in estimation and hypothesis testing, which are discussed in Chapter 6.

The Normal Approximation to the Binomial Distribution   As shown previously, when *n* is large (beyond the binomial tables for *n*) and *p* is small (such that *np* ≤ 5), the binomial distribution can be approximated with the Poisson distribution. When *n* is beyond the table values and *p* is neither very large nor very small, the binomial distribution can be approx-

TABLE 4-8   Commonly used z values

Configuration of area	Percentage of area	z value
(normal curve, shaded between $-z$ and $+z$)	.90	±1.64
	.95	±1.96
	.99	±2.58
(normal curve, shaded to right of $-z$)	.90	−1.28
	.95	−1.64
	.99	−2.33
(normal curve, shaded to left of z)	.90	1.28
	.95	1.64
	.99	2.33

imated with the normal distribution. The guidelines for using the normal approximation to the binomial are

$$n \geqslant 30$$
$$np \text{ and } n(1 - p) \geqslant 5$$

To demonstrate the approximation, consider the probability that 50 jurors selected randomly from a jury pool that contains 55 percent women will include 30 or more women. The selection is a binomial process with $n = 50$, $p = .55$, and $x \geqslant 30$. The binomial tables do not include an n of 50, and the Poisson approximation is inappropriate because $np = 27.5$, which is greater than 5. [The technique of using $1 - p$ for p will not work in this case either, since $n(1 - p) = 23.5$.] But the criteria for a normal approximation are met and the binomial equivalents of the normal parameters are

$$\mu_B = np = 27.5$$
$$\sigma_B = \sqrt{np(1 - p)} = 3.52$$

Before determining the normal probability $P(X \geqslant 30|\mu = 27.5, \sigma = 3.52)$, a correction must be made to permit the approximation of a discrete distribution with a continuous distribution. (Recall that the probability of a discrete value in a continuous distribution is zero.) In continuous data, we define 29 as "28.5 to 29.5," 30 as "29.5 to 30.5," 31 as "30.5 to 31.5," and so on, thus giving area to values that previously had none. The discrete values equal to or greater than 30 are now represented by the area to the right of 29.5, as shown in Figure 4-9.

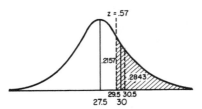

Figure 4-9 Normal approximation to the binomial distribution.

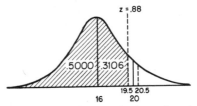

Figure 4-10 Normal approximation to the Poisson distribution.

The problem is now solved as a regular normal distribution problem:

$$z = \frac{29.5 - 27.5}{3.52} = .57$$

$$P(X \geq 29.5 | \mu = 27.5, \sigma = 3.52) = .5000 - .2157 = .2843$$

The approximate probability[7] that 30 or more jurors will be women is .2843.

The Normal Approximation to the Poisson Distribution When the mean of a Poisson process is greater than 10, it is not possible to use the table of Poisson probabilities. The normal approximation to the binomial distribution can be extended to the Poisson distribution in such cases, the only criterion being that $\lambda > 10$. For example, if the mean number of automobiles arriving at a service station between 4 and 5 P.M. on weekdays is 16, what is the probability that fewer than 20 will arrive during this period next Tuesday?

The Poisson mean λ is more than 10, so the normal approximation can be used. The Poisson equivalents of the normal parameter are

$$\mu_P = \lambda = 16$$
$$\sigma_P = \sqrt{\lambda} = 4$$

The correction from the discrete to continuous distribution requires "fewer than 20" to be defined as "less than or equal to 19.5." The area corresponding to the normal probability $P(X \leq 19.5 | \mu = 16, \sigma = 4)$ is shown in Figure 4-10 and is computed below:

$$z = \frac{19.5 - 16}{4} = .88$$

$$P(X \leq 19.5 | \mu = 16, \sigma = 4) = .5000 + .3106 = .8106$$

[7]The true binomial probability, $P(X \geq 30 | n = 50, p = .55)$ is .2862.

The approximate probability[8] of fewer than 20 automobiles, then, is .8106.

PROPORTIONS

An absolute number can be converted to a percentage or *proportion* simply by dividing it by a total. Thus, instead of saying that 48 of a sample of 120 employees are women, we might say that the proportion of women is 48/120 = .40.

When the process or characteristic measured by a proportion follows a binomial distribution, the same expedient of dividing by the total can be used to convert the binomial distribution into a distribution of proportions. In the example above, the binomial parameters are

$$n = 120 \qquad p = .40$$

and the binomial mean and standard deviation are

$$\mu_B = np = (120)(.40) = 48$$
$$\sigma_B = \sqrt{np(1 - p)} = \sqrt{(120)(.40)(.60)} = \sqrt{28.8} = 5.37$$

To find the mean and standard deviation of the corresponding distribution of proportions, we again divide by n

$$\text{Mean} = \frac{\mu_B}{n} = \frac{np}{n} = p = .40$$
$$\text{Standard deviation} = \frac{\sigma_B}{n} = \frac{\sqrt{np(1 - p)}}{n} = \sqrt{\frac{np(1 - p)}{n^2}} = \sqrt{\frac{p(1 - p)}{n}}$$
$$= \sqrt{\frac{(.4)(.6)}{120}} = \sqrt{.002} = .045$$

For population data, the usual notation is

$$\text{Mean proportion} = \pi \qquad (4\text{-}8)$$

$$\text{Standard error of proportion} = \sigma_\pi = \sqrt{\frac{\pi(1 - \pi)}{n}} \qquad (4\text{-}9)$$

---

[8]The exact Poisson probability of $P(X < 20|\lambda = 16)$ is .8122.

The use of π as the symbol for the population proportion should not be confused with the constant π that is used to find the area and diameter of circles.[9]

The distribution of proportions is really a form of the normal approximation to the binomial and, as such, is appropriate only when n is equal to or greater than 30 and np and $n(1 - p)$ are both equal to or greater than 5. Since the distribution of proportions is assumed to be normal, there is no need for a table of probabilities of proportions; we merely use the normal distribution with a mean of π and a standard deviation of σ_π. Values of z for such a distribution are found by

$$z = \frac{p - \pi}{\sigma_\pi} \qquad (4\text{-}10)$$

where p is a random variable in proportional form, corresponding to X in formula (4-7), and π and σ_π are the counterparts of μ and σ in the same formula.

Applications of the distribution of proportions always involve sampling, which is discussed in Chapter 5. Since this is an integrative application involving several statistical techniques, we defer an illustration of the distribution of proportions to Chapter 14.

A MANAGER'S PERSPECTIVE

The preceding material may seem more appropriate for a mathematician or a theoretical statistician than for a practicing manager. Indeed, most managers do not, on a routine basis, deal with distributions of data. However, managers do base decisions upon data which follow theoretical probability distributions. Where is the linkage between probability theory and decision making? The "missing link," as we might call it, is the *samples* of managerial data on which many decisions are based. In the next chapter we shall see how these samples are obtained and how they relate to theoretical probability distributions.

---

[9]The constant π has a value of 3.14159 while the proportion π is a variable and can assume any value between 0 and 1.

CHAPTER 5

Sampling and Sampling Distributions

Population data are often impossible, impracticable, or too costly to collect. For example, we cannot conduct a census of the population of cubic meters of air to determine pollution indexes. It is not practicable to determine the mean life of a shipment of light bulbs by census for the simple reason that the test process will burn out all the bulbs. The success of a new product could be predicted by a census of all consumers in the market area, but the cost of such a census would probably exceed the profit from sales. In cases like these, a portion of the population, called a *sample*, can be measured and the sample statistics used to draw inferences about the parent population.

Sampling is a powerful statistical tool and, like many powerful tools, it must be thoroughly understood and used with caution if the results are to be beneficial. It is all too easy to call a few handy items a sample and to draw sweeping conclusions about the population. Only if the sample is selected according to carefully designed plans and the results suitably qualified can statistical inference be beneficial to managers. The development of a sampling plan is itself a matter of specialized study, but we can cover a few of the highlights here.

SAMPLING METHODS

The most important quality of a sample is the extent to which it is representative of the population. A representative sample will reflect the various patterns and subclasses of the population as well as provide good estimates of the population parameters. Representativeness can be achieved best by *random* selection methods, which give every item in the population an equal opportunity to appear in the sample.

Simple Random Sampling

When a population is relatively homogeneous, that is, variance in the characteristic under study is due to probability rather than innate differences between subpopulations, representativeness can be achieved by *simple* random sampling. In this sense, "simple" means that the sample is drawn from a single list, or *frame,* consisting of the entire population.

There are a number of ways to achieve randomness in a sample. Drawing numbers from a hat is reasonably random, as are the various means for drawing numbers in a bingo game. In statistics and scientific testing, the use of random numbers that can be related to specific items in the frame is a common technique. Random numbers are particularly convenient in computer-assisted methods, since some programming languages include the capability for generating random numbers internally. Tables of random numbers, like the one in Appendix H, are readily available for manual use.

To illustrate the use of random numbers in a simple random sample, let us assume that a sample of hand-held calculators is to be drawn from yesterday's run of Little Wizards at the Black Box Calculator Company.[1] The calculators are numbered serially from 664001 to 664500. Since the first three digits in each serial number are the same, they can be ignored and each calculator can be identified by the last three digits. Although there may be no physical listing of calculators from which to draw the sample, the numbers 001 to 500 uniquely identify the 500 calculators and will serve quite nicely as a frame.

The serial numbers of calculators to be included in the sample are taken from the table of random numbers, which is partially reproduced as Table 5-1. Although the random numbers consist of five digits, only three are necessary to identify a serial number. They can be the first three, the last three, the first two and the fourth, or any combination of three digits that is convenient. The designation of digits to be used should be consistent from number to number, however. For this example, we shall use the first three digits.

The first calculator in the sample should be serial number 100, since

---

[1] Since all calculators are of the same make and model, it is safe to assume homogeneity. This might not be true if three or four models with different characteristics were included in the population.

TABLE 5-1   Partial table of random numbers*

10097	66065	98520	65481	85017
37542	31060	11805	80124	16719
08422	85269	83452	74350	65842
99019	63573	88685	69916	76875
12807	73796	99594	09893	93640

*From Appendix H.

the first random number is 10097. (The first random number can be *any* number in the table, but subsequent numbers should then be selected in order.) The second calculator is number 375, the third, 84, and so on. Numbers greater than 500, such as the 990 of the fourth random number, are simply ignored. In this manner, any limits can be placed on the range of random numbers, and numbers outside the limits will not be considered. The numbers within the limits are still distributed randomly.

Stratified Random Sampling

When the population is not homogeneous, or when a single population frame is not available, *stratified random sampling* may be appropriate. A sample is stratified when the selection process provides items from the population in some predetermined ratio. For example, assume that a sample of postal employees is to be taken to determine the mean length of employment. Because women have entered the work force in large numbers only recently, their mean length of employment is likely to be less than that of men. Thus the population is not homogeneous with respect to length of employment. Furthermore, if men outnumber women 4 to 1 in the Postal Service, we may wish to reflect this ratio in the sample.[2] A stratified random sample could be drawn according to the following rule:

If the first digit of the random number is a 0 or 1, select from the frame of female employees; if the first digit is 2 to 9, select from the frame of male employees.

[2] A sample that reflects the same ratio between strata as the population is a *proportional* stratified sample. There are special cases in which a *nonproportional* stratified sample is appropriate, for example, to ensure some minimum representation of each stratum or to give weight to a stratum judged more important than the others.

Now assume that the frame of male employees has 68,800 names and the frame of female employees has 17,200. The sample selection process would proceed as follows:

Step	Random number	Action
1	10097	Select frame of female employees
2	37542	Omit; too large for frame
3	08422	Select 8422d female employee
4	99019	Select frame of male employees
5	12807	Select 12,807th male employee
6	66065	Select frame of male employees
7	31060	Select 31,060th male employee
Last		Continue until desired sample size is reached

Nonrandom Sampling

When time and cost factors limit the collection of a sample, or when only a rough estimate of a population parameter is desired, a nonrandom sampling technique may be appropriate. A few of the more common nonrandom techniques are explained briefly.

Systematic Sampling   A systematic sample consists of items found at some regular interval in the frame. For example, a systematic sample of a production process might contain every 20th item as it is produced. There is a slightly random element in this process (the selection of the first item), but the systematic selection may not ensure representativeness. To illustrate this point, assume that a final assembly line is fed, alternately, by five identical subassembly machines, the 5th one of which is badly out of tolerance. Thus, the 5th, 10th, 15th, 20th, etc., items will be defective. Now, if the sample consists of the 1st, 21st, 41st, 61st, etc., items, a defective item will never appear in the sample when, in fact, the population is 20 percent defective!

Judgment Sampling   When testing is expensive (as in crashing automobiles to test air-bag efficiency), a necessarily small sample may not be representative if it is chosen randomly. For example, if 10 automobiles were selected randomly from a list of all makes and models, the sample might contain seven compacts and three full-sized cars from only two of the 15 or 20 major manufacturers. In this case, it would be better to use a *judgment sample* in which the researcher subjectively selects 10 automobiles he or she considers representative of the population. The judgmentally selected sample might include 10 different makes, some of every size, a few station wagons and sports cars, and so on.

The danger of judgment sampling is that the person selecting the sample may, knowingly or not, introduce personal biases into the selection process. The quality control inspector who genuinely believes his company's product is of superior quality may unwittingly select for inspection the output of the most skilled employee. School administrators may select as "representative" their best students to participate in a testing program. And consumer advocates are often accused of using only the worst cases to represent typical quality in products. Judgment sampling can achieve representativeness in a much smaller sample than random sampling can, but it must be used with great caution to prevent bias.

Quota Sampling   A variation of judgment sampling in which items are selected in predetermined quantities from subclasses of the population is called *quota sampling*. In this sense, quota sampling is a form of stratified judgment sampling.[3] To continue the automobile example, if 20 percent of the automobile population were station wagons, 2 of the 10 automobiles selected for a quota sample would be station wagons, the exact make and size of the station wagon being left to the judgment of the person collecting the sample.

Like stratified random sampling, quota sampling is appropriate when the population is not homogeneous. Quota sampling also may minimize bias in judgment by restricting the number of items that can be chosen from any specific category.

Convenience Sampling   As the name suggests, the criteria for inclusion in a *convenience sample* are merely that the items are easy to identify and convenient to collect. For obvious reasons, the use of convenience sampling is widespread, perhaps more so than we like to admit. A person asked to collect data for an opinion poll may find it convenient to interview her friends (who simply because they are her friends are likely to have similar tastes). Or an interviewer may collect data at his favorite bar, where the patrons share similar economic, sociological, and ethnic backgrounds. It may also be convenient to use one case of 48 spark plugs as a sample of the daily output of a spark plug factory. But the contents of a single case probably were produced within a relatively short interval. If that interval occurred immediately after setting and calibrating the machinery used in production, the chance of substandard spark plugs is much less than if the interval occurred just before the machinery was due to be recalibrated.

---

[3]A metaphor suggested by Robert Parsons, *Statistical Analysis: A Decision-Making Approach,* Harper & Row, New York, 1974, which is an excellent, comprehensive statistics text.

In these cases, or in any convenience sample for that matter, the representativeness of the sample is suspect. Indeed, the term "convenience sample" is sometimes used in a pejorative sense to ridicule what is believed to be an unscientific sample. A convenience sample should be used only when the population is very homogeneous or the need for immediacy is so great that no other sampling technique can be used. In the latter case, it is suggested that the results of the convenience sample be confirmed with another, perhaps random sample, when time permits.

SAMPLE STATISTICS

The counterparts of population parameters are called *statistics* when the data set is a sample. Although parameters and statistics share common names and definitions, there are slight differences in notation and computation that should be noted before discussing sampling distributions.

The Sample Mean

The counterpart of the population mean μ is the sample mean $\overline{X}$, read "X bar." Like the population mean, the sample mean is found by dividing the sum of values by the number of values. However, a lowercase n is used to distinguish the number of values in the sample from N, the number of values in the population:

$$\text{Sample mean} = \overline{X} = \frac{\Sigma X}{n}$$

The use of n as a sample size is consistent with binomial notation, where n is the number of trials. In both cases, n is really a sample size, since the number of, say, tosses of a coin, is only a sample of the infinite number of coin tosses that makes up the population.

Table 5-2 shows a population of hourly wage rates in a factory, arranged in order of employee social security number, in which a systematic sample (every tenth value) is identified by an asterisk. For comparison, both the population and sample means have been computed in Table 5-2.

The Sample Standard Deviation

The sample standard deviation s is computed like the population standard deviation σ except that the denominator is *1 less* than the number of items and $\overline{X}$ and n are used in place of μ and N

TABLE 5-2   Population and sample means of hourly wage rates in a factory in dollars

i	X	i	X	i	X	i	X	i	X
1	4.00	21	3.50	41	4.00	61	3.00	81	3.00
2	3.25	22	3.00	42	3.50	62	3.50	82	3.75
3	4.50	23	2.75	43	2.75	63	5.00	83	4.50
4	3.75	24	6.00	44	3.25	64	3.00	84	3.50
5	3.75	25	4.50	45	1.00	65	3.75	85	3.75
6	3.50	26	3.75	46	6.00	66	4.50	86	3.25
7	5.00	27	3.25	47	3.75	67	3.50	87	3.50
8	3.50	28	4.00	48	3.00	68	4.00	88	2.75
9	3.75	29	3.75	49	3.50	69	3.50	89	6.00
10	2.75*	30	3.50*	50	3.50*	70	5.00*	90	4.50*
11	3.00	31	3.00	51	3.25	71	3.25	91	3.75
12	3.25	32	2.75	52	4.50	72	3.75	92	3.50
13	3.75	33	5.00	53	4.00	73	4.00	93	3.75
14	4.00	34	3.50	54	3.00	74	2.75	94	4.00
15	2.75	35	3.00	55	3.50	75	3.75	95	3.00
16	3.50	36	3.75	56	3.00	76	6.00	96	4.00
17	5.00	37	2.75	57	3.75	77	3.50	97	4.50
18	3.00	38	3.50	58	4.50	78	3.50	98	3.50
19	2.75	39	4.50	59	6.00	79	3.00	99	3.00
20	4.00*	40	3.00*	60	2.75*	80	4.50	100	4.00*

$$\text{Population mean} = \mu = \frac{\Sigma X}{N} = \frac{371.25}{100} = 3.7125$$

$$\text{Sample mean} = \overline{X} = \frac{\Sigma X}{n} = \frac{37.50}{10} = 3.750$$

* Every tenth value.

$$\text{Sample standard deviation} = s = \sqrt{\frac{\Sigma(X - \overline{X})^2}{n - 1}} \qquad (5\text{-}2)$$

$$\text{Shortcut formula: } s = \sqrt{\frac{\Sigma X^2 - (\Sigma X)^2/n}{n - 1}} \qquad (5\text{-}3)$$

The use of $n - 1$ in computing s has been found to give a better estimate of σ than computation with n does. The complete justification of $n - 1$ is beyond the scope of this book, but it may be helpful to know that the denominator in this formula represents *degrees of freedom* and is reduced by 1 because one term, $\overline{X}$, is itself an estimate. Later, in linear-regression analysis, we shall estimate two parameters and incorporate $n - 2$ degrees of freedom into a formula.

The standard deviation of the systematic sample of hourly wage rates

TABLE 5-3   Mean and standard deviation of a systematic sample of hourly wage rates in dollars

i	X	$\overline{X}$	$X - \overline{X}$	$(X - \overline{X})^2$
10	2.75	3.75	−1.00	1.0000
20	4.00	3.75	.25	.0625
30	3.50	3.75	−.25	.0625
40	3.00	3.75	−.75	.5625
50	3.50	3.75	−.25	.0625
60	2.75	3.75	−1.00	1.0000
70	5.00	3.75	1.25	1.5625
80	4.50	3.75	.75	.5625
90	4.50	3.75	.75	.5625
100	4.00	3.75	.25	.0625
	$\Sigma X = 37.50$			$\Sigma(X - \overline{X})^2 = 5.5000$

$$\overline{X} = \frac{\Sigma X}{n} = \frac{37.50}{10} = 3.75 \qquad s = \sqrt{\frac{\Sigma(X - \overline{X})^2}{n - 1}} = \sqrt{\frac{5.5}{9}} = \sqrt{.6111} = .7817$$

is computed in Table 5-3. The sample standard deviation of .7817 compares most favorably with the population standard deviation of .7837, the computation of which is omitted here because of its length. The ambitious reader may confirm the population standard deviation using the data in Table 5-2 and formula (3-5) or (3-6). The authors, who are not quite that ambitious, determined the population standard deviation with the aid of a scientific calculator.

That the sample standard deviation should be accurate to two decimal places is a bit lucky and should not be expected in all cases. For example, Table 5-4 shows a sample selected with random numbers using the first two digits of the numbers in Table 5-1 (where $01 = 1, 02 = 2, \ldots, 00 = 100$). This sample of 10 has a mean of \$3.60 and a standard deviation of \$.7188. Other samples will show similar deviations from the population mean of \$3.7125 and standard deviation of \$.7837.

SAMPLING DISTRIBUTIONS

The samples illustrated in Tables 5-3 and 5-4 are but two of a very large number of samples of size 10 that can be drawn from the population of 100 hourly wages.[4] The means of these samples constitute a distribution

---

[4]"Very large" is an inadequate term to describe the true size of this number. There are over 17 *trillion* (17,000,000,000,000) different samples of size 10 that can be drawn from a population of 100!

TABLE 5-4   Mean and standard deviation of a random sample of hourly wage rates in dollars

RN i	Wage X	$X - \overline{X}$	$(X - \overline{X})^2$
10	2.75	.85	.7225
37	3.25	.35	.1225
08	3.50	.10	.0100
99	3.00	.60	.3600
12	3.25	.35	.1225
66	4.50	−.90	.8100
31	3.00	.60	.3600
85	3.75	−.15	.0225
63	5.00	−1.40	1.9600
73	4.00	−.40	.1600
	$\Sigma X = 36.00$		$\Sigma(X - \overline{X})^2 = 4.6500$

$$\overline{X} = \frac{\Sigma X}{n} = \frac{36.00}{10} = 3.60 \qquad s = \sqrt{\frac{\Sigma(X - \overline{X})^2}{n - 1}} = \sqrt{\frac{4.65}{9}} = \sqrt{.5167} = .7188$$

in much the same way that the original 100 items make up a distribution. A distribution of sample means, called a *sampling distribution,* is a continuous probability distribution defined by two parameters, the mean and the standard deviation. The mean and standard deviation will be recalled as the parameters of another theoretical probability distribution, the normal distribution. While the mean and standard deviation of a sampling distribution can be computed similarly to their normal distribution counterparts, there are shortcuts facilitated by the close relationship between a sampling distribution and its parent distribution, the population.

The Mean of a Sampling Distribution

The mean of a sampling distribution cannot be demonstrated with any but a very small population and even smaller samples. Table 5-5 shows such a small population ($N = 5$) and the distribution of the 10 possible samples of size 3 ($n = 3$). When the means of the samples $\overline{X}$ are treated as individual items in a distribution of $N_{\overline{X}}$ means, the *mean of the means* $\mu_{\overline{X}}$ can be found in the conventional manner. The computations in Table 5-5 show that $\mu_{\overline{X}}$ is equal to μ and either term can be used to represent the *mean of the sampling distribution.*

The Standard Error of the Mean

The standard deviation of a sampling distribution is called the *standard error of the mean* $\sigma_{\overline{X}}$ and can also be computed in the conventional manner

TABLE 5-5 *The mean of a sampling distribution for population {1, 2, 3, 4, 5}*

Sample number, i	Sample, $n = 3$	Sample mean $\overline{X} = \dfrac{\Sigma X}{n}$
1	1, 2, 3	2
2	1, 2, 4	$2\frac{1}{3}$
3	1, 2, 5	$2\frac{2}{3}$
4	1, 3, 4	$2\frac{2}{3}$
5	1, 3, 5	3
6	2, 3, 4	3
7	1, 4, 5	$3\frac{1}{3}$
8	2, 3, 5	$3\frac{1}{3}$
9	2, 4, 5	$3\frac{2}{3}$
10	3, 4, 5	4
		$\Sigma\overline{X} = 30$

$$\text{Mean of population} = \mu = \frac{\Sigma X}{N} = \frac{1 + 2 + 3 + 4 + 5}{5} = \frac{15}{5} = 3$$

$$\text{Mean of sampling distribution} = \mu_{\overline{X}} = \frac{\Sigma\overline{X}}{N_X} = \frac{2 + 2\frac{1}{3} + 2\frac{2}{3} + \cdots + 4}{10} = \frac{30}{10} = 3$$

$$\mu = \mu_X$$

if the distribution is sufficiently small. The standard deviation for the means of the samples taken from the population is computed in Table 5-6.

The mean of the sampling distribution is conveniently equal to the mean of the population, but the relationship between the standard error

TABLE 5-6 *The standard deviation of a sampling distribution from sample means*

i	$\overline{X}$	$\mu_{\overline{X}}$	$\overline{X} - \mu_{\overline{X}}$	$(X - \mu_{\overline{X}})^2$
1	2	3	-1 or $-\frac{3}{3}$	$\frac{9}{9}$
2	$2\frac{1}{3}$	3	$-\frac{2}{3}$	$\frac{4}{9}$
3	$2\frac{2}{3}$	3	$-\frac{1}{3}$	$\frac{1}{9}$
4	$2\frac{2}{3}$	3	$-\frac{1}{3}$	$\frac{1}{9}$
5	3	3	$()$	0
6	3	3	0	0
7	$3\frac{1}{3}$	3	$\frac{1}{3}$	$\frac{1}{9}$
8	$3\frac{1}{3}$	3	$\frac{1}{3}$	$\frac{1}{9}$
9	$3\frac{2}{3}$	3	$\frac{2}{3}$	$\frac{4}{9}$
10	4	3	1 or $\frac{3}{3}$	$\frac{9}{9}$
				$\Sigma(\overline{X} - \mu_{\overline{X}})^2 = 30/9$

$$\text{Standard error of the mean} = \sigma_{\overline{X}} = \sqrt{\frac{\Sigma(\overline{X} - \mu_{\overline{X}})^2}{N_{\overline{X}}}} = \sqrt{\frac{30/9}{10}} = \sqrt{1/3} = .5774$$

of the mean and the population standard deviation is not as simple. The precise relationship is represented by

$$\sigma_{\bar{x}} = \frac{\sigma}{\sqrt{n}} \sqrt{\frac{N-n}{N-1}} \qquad (5\text{-}4)$$

The second expression in this formula, $\sqrt{(N-n)/(N-1)}$, called the *finite-population correction,* is necessary only when the sample is a significant portion of the population. The rule of thumb generally applied is that the finite correction factor can be dropped when the sample is less than 5 percent of the population. At this point the correction is approximately .98 and is considered close enough to 1.00 to be ignored.[5]

Since most business applications deal with large populations, we rarely heed the finite-population correction and can use the simpler formula

$$\sigma_{\bar{x}} = \frac{\sigma}{\sqrt{n}} \qquad (5\text{-}5)$$

For the distribution of samples of 3 taken from the population of 5, n/N is quite large, and equation (5-4) must be used. This computation, shown in Table 5-7, results in exactly the same value for $\sigma_{\bar{x}}$ determined in Table 5-6.

TABLE 5-7   The standard deviation of a sampling distribution from the population standard deviation

i	X	μ	$X - \mu$	$(X - \mu)^2$
1	1	3	-2	4
2	2	3	1	1
3	3	3	0	0
4	4	3	1	1
5	5	3	2	4
				$\Sigma(X - \mu)^2 = 10$

$$\sigma = \sqrt{\frac{\Sigma(X-\mu)^2}{N}} = \sqrt{\frac{10}{5}} = \sqrt{2}$$

$$\sigma_{\bar{x}} = \frac{\sigma}{\sqrt{n}} \sqrt{\frac{N-n}{N-1}} = \frac{\sqrt{2}}{\sqrt{3}} \sqrt{\frac{5-3}{5-1}} = (.8165)(.7071) = .5774$$

[5] For $N = 100$ and $n = 5$, the finite correction factor is $\sqrt{95/99}$, or .9796. For smaller values of n (or larger values of N), n/N is less than .05 and the finite correction becomes even closer to 1.

Theoretical Sampling Distributions

Distributions of sample means follow one of two theoretical probability distributions based on the size of the samples and the characteristics of the population from which the samples are drawn. To avoid the confusion that frequently arises when every combination of size and characteristics is considered, regardless of how rarely it may occur, we limit this discussion to situations common in managerial decision making.

Large Samples   The theoretical distribution for the mean of large samples, regardless of the distribution of the parent population, is the normal distribution. This fact is known in statistics as the *Central-Limit Theorem* which states, more formally, that

The sample mean $\overline{X}$ approaches a normally distributed random variable with a mean of μ and a standard deviation of $\sigma/\sqrt{n}$ as the sample size n increases.

Thus, by the simple expedient of using a large sample, a nonnormal or undefined population can be converted into a normally distributed sampling distribution. For example, the uniformly distributed population in Figure 5-1 (represented by the rectangle) can be thought of as a sampling distribution of sample size 1. As sample size increases, the distribution of sample means approaches normality. For large samples, the means can be

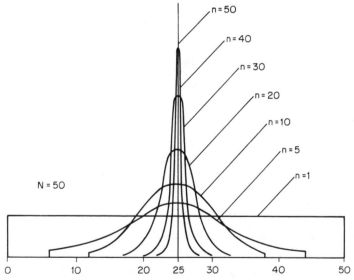

Figure 5-1   Sampling distributions from a uniform population.

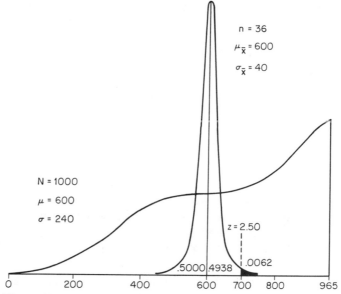

Figure 5-2 Sampling distribution of FICA taxes.

considered to follow a normal distribution, a very peaked, or *kurtotic,* normal distribution because of the scale used in Figure 5-1 but a normal distribution none the less. Of course, when n finally equals N, the sample mean can never be anything but the population mean and the "distribution" is a vertical line at the mean.

In ordinary usage, "large" is a rather vague term. In reference to statistical sample size, however, "large" has a definite meaning. A sample is considered to be large when $n \geqslant 30$.

To illustrate an application of the Central-Limit Theorem, consider the distribution, shown in Figure 5-2, of annual FICA taxes for 1000 employees. The distribution has a mean of $600 and a standard deviation of $240, but the upper limit of $965 on FICA taxes gives the curve an unusual shape.

It is impossible to find the probability that the FICA taxes for any one employee will exceed a given amount, say, $700, because there is no table of probabilities for such a distribution. However, the probability that the mean of a sample of 36 employees will exceed $700 can easily be computed because the sampling distribution for $n = 36$ is normal. As for any normal distribution, the area under a portion of the curve can be found in the table for the standard normal distribution if the distance from the mean is converted into a z value. The value of z is computed in the usual manner with changes only to acknowledge that this is a distribution of $\bar{X}$'s, not X's:

$$z = \frac{X - \mu}{\sigma} = \frac{\bar{X} - \mu_{\bar{X}}}{\sigma_{\bar{X}}}$$

$\mu = 600 \qquad \sigma = 240 \qquad \bar{X} = 700 \qquad N = 1000 \qquad n = 36$

$$\frac{n}{N} = \frac{36}{1000} = .036 < .05 \text{ (no finite population correction)}$$

$$\sigma_{\bar{X}} = \frac{\sigma}{\sqrt{n}} = \frac{240}{\sqrt{36}} = 40$$

$$z = \frac{\bar{X} - \mu_{\bar{X}}}{\sigma_{\bar{X}}} = \frac{700 - 600}{40} = 2.50$$

Since 700 is 2.50 standard deviations away from 600, the probability that a sample mean lies between 600 and 700 is .4938 (from Appendix F) and the probability that a sample mean is greater than \$700 is .5000 − .4938 = .0062.

Of course, if the parent population is normal, the distribution of large sample means is also normal. The distribution of means of samples of size 36 taken from a normal population of 500 items with a mean of 350 and a standard deviation of 56 is shown in Figure 5-3.

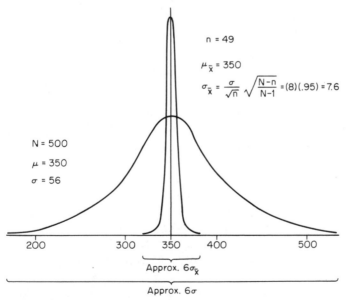

$$n = 49$$
$$\mu_{\bar{x}} = 350$$
$$\sigma_{\bar{x}} = \frac{\sigma}{\sqrt{n}} \sqrt{\frac{N-n}{N-1}} = (8)(.95) = 7.6$$

$N = 500$
$\mu = 350$
$\sigma = 56$

200   300   350   400   500

Approx. $6\sigma_{\bar{x}}$

Approx. 6σ

Figure 5-3 Sampling distribution from a normal population.

Small Samples The distribution of means of small samples follows a known theoretical distribution *only when the parent population is normal.* Within this restriction, there are two possible theoretical sampling distributions, the choice being based on knowledge of the population standard deviation.

Known Population Standard Deviation. When the parent population is normal and the population standard deviation is known, the sampling distribution is *normal.* Areas of a normal sampling distribution of small samples are found in exactly the same manner as areas of a normal sampling distribution of large samples.

Unknown Population Standard Deviation. When the population standard deviation is unknown and must be estimated, additional variance is introduced into the sampling distribution. The resultant curve is flatter and has longer tails than a normal curve. Moreover, this flattening and elongation of tails becomes more pronounced as sample size decreases. Thus, unlike normal distributions which can be described by a single, standard normal distribution, there is a different small-sample distribution for each small sample size.

The *t* Distribution The theoretical distribution that describes small-sample distributions is called the *t*, or *Student's, distribution,* after the pen name of W. M. Gosset, a statistician of the early 1900s. Gosset developed the *t* distribution as a result of his work with small samples in a New York brewery. We shall not speculate on the heights to which he might have risen had his work involved *large* samples.

Since the *t* distribution explicitly requires an estimate of the population standard deviation, we defer illustrative examples until Chapter 6, which covers the estimation process. It is appropriate at this point, however, to examine the procedures for using tables of the *t* distribution.

The table of probabilities of the standard normal distribution is entered with z, the number of standard deviations from the mean, and yields the probability that X lies not more than z standard deviations from the mean. The table of t values is entered with two arguments (values that determine the appropriate row and column), namely, degrees of freedom (df) and the proportion of area in one tail, to yield t, the number of standard deviations between the tail and the mean.

The argument for degrees of freedom establishes which of the many t distributions (one for each sample size) is to be used. The expression of this argument in degrees of freedom, $n - 1$, rather than sample size, n, is consistent with the use of $n - 1$ in the computation of s, the standard deviation of the sample mean. Later, we shall see that s is routinely used with the t distribution in estimation and hypothesis testing.

The argument in terms of tail area is also based upon typical applications of the t distribution in estimation and hypothesis testing.

We can now compare, in Figure 5-4, the shapes of various t distributions with the normal distribution. Notice that given tail areas, .025 on the right and .05 on the left, begin at points more distant from the mean as sample size decreases. The distance from the mean is expressed in standard deviations, z or t. The reader should confirm these t values in

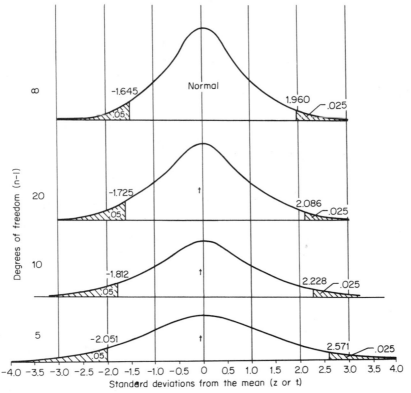

Figure 5-4   Comparison of normal and t distributions.

Appendix G to ensure familiarity with the table before proceeding to Chapter 6.

Although sampling distributions can always be described by one of two distributions, the normal or t, the circumstances under which each distribution applies can be confusing. Table 5-8 provides a summary of these circumstances that may be useful for later applications.

TABLE 5-8   *Theoretical sampling distributions*

Sample size n	Population	Population standard deviation σ	Sampling distribution
$\geqslant 30$	Any	Known	Normal
		Unknown*	Normal
<30	Normal	Known	Normal
		Unknown*	t

*Estimated from sample standard deviation s.

This chapter completes the theoretical background necessary for two of the most important statistical methods, estimation and hypothesis testing. While the material in these last three chapters is important, it certainly is not necessary to master it for occasional use. Instead, we suggest that the reader refer back to these chapters while reading Chapters 6 and 7 and, in a few instances, the chapters on quantitative methods in Part 3.

CHAPTER 6

Statistical Inference

In Chapter 5 population data were used to determine information about a sampling distribution. The *deductive* process of applying population data to samples is a useful statistical technique, but it has an obvious limitation: the population data must be known. When population data are not known, as is usually the case, an *inductive* process of generalizing from sample data is followed. Inductive reasoning in statistics, usually referred to as *statistical inference,* includes two fundamental processes, *estimation* and *hypothesis testing.* Both processes have direct applications to managerial decision making and deserve a detailed examination.

ESTIMATION

In common usage, an estimate is a guess or an opinion. A golfer estimates the distance from his ball to the green, a witness to an accident may be asked to estimate the speed of an automobile, and a grocer estimates the number of watermelons he could sell at $1 each. These are *subjective* estimates, based on the estimator's experience, judgment, and competence. As subjective estimates, they are roughly analogous to the subjective probabilities discussed in Chapter 4.

In statistical usage, estimates are more analogous to *empirical* probabilities; that is, they are based on observed data. Furthermore, to serve a statistical purpose, these estimates must convey useful statistical information. When population data are unknown or unattainable, the most useful estimates are those of the population parameters, the features that uniquely define the population. Although *any* population parameter can be estimated, in this chapter we focus on the parameters of the normal distribution, the mean and the standard deviation.

113

Point Estimates

An estimate that consists of a single value is called a *point estimate*. In statistical notation, the estimate of a parameter is identified with the hat symbol.[1] Thus, the estimate of the population mean μ is $\hat{\mu}$ and the estimate of the population standard deviation σ is $\hat{\sigma}$.

Point estimates of population parameters are based on the corresponding sample statistics, which are called *estimators* when they are used for this purpose. Thus, the sample mean $\overline{X}$ is the estimator for the point estimate of the population mean μ, and the sample standard deviation s is the estimator for the point estimate of the population standard deviation σ. The relationship between estimators, estimates, and parameters is summarized in Table 6-1.

TABLE 6-1   Parameters, estimators, and point estimates of the normal distribution

	Parameter	Estimator	Estimate
Mean	μ	$\overline{X}$	$\hat{\mu}$
Standard deviation	σ	s	$\hat{\sigma}$

To illustrate the development of point estimates, assume that a committee of concerned consumers in Atlanta wants an estimate of the mean price for 1-pound cans of coffee in their city. They decide to take a quota sample (several selected brands from each of a number of independent and chain grocery stores) and gather the data shown in Table 6-2.

The sample data have a mean of $3.07 and a standard deviation of $.18. These sample statistics are the estimators of the corresponding population parameters and can be used in lieu of the parameters. For example, the computation of the cost of a typical "market basket" that includes 1 pound of coffee would use a coffee price of $3.07.

Interval Estimates

When an estimate is used in further computations, as the coffee price is used in computing a market-basket price, the single value of the point estimate is most convenient. However, we know that the true mean price of coffee probably is not exactly $3.07 and another sample most likely would lead to a different estimate. How confident can the consumer group be in the accuracy of the estimate of $3.07?

Confidence in an estimate is expressed as the probability that the population mean will fall in a certain *interval* about the point estimate. The

[1]The symbol ˆ is one of several phonetic marks called *circumflexes*, but it is commonly referred to as a "hat" in statistics. One student, whose sense of humor far exceeded his skill in statistics, observed that the symbol bears a certain resemblance to a *dunce* hat.

TABLE 6-2   Sample data of coffee prices

i	Price X	i	Price X	i	Price X
1	$3.07	11	$3.03	21	$2.95
2	3.06	12	3.19	22	3.38
3	2.92	13	3.15	23	3.36
4	2.91	14	2.90	24	2.94
5	3.00	15	3.14	25	3.13
6	3.29	16	3.30	26	3.21
7	3.18	17	3.22	27	2.78
8	3.14	18	2.84	28	2.80
9	2.91	19	3.07	29	3.22
10	2.96	20	2.73	30	3.41

$$\hat{\mu} = \overline{X} = \frac{\Sigma X}{n} = \frac{92.19}{30} = 3.07$$

$$\hat{\sigma} = s = \sqrt{\frac{\Sigma X^2 - (\Sigma X)^2/n}{n-1}} = \sqrt{\frac{284.2857 - 283.30}{29}} = .18$$

upper and lower limits of the interval are determined by three factors: the confidence level of the estimate, the sample mean, and the standard error of the mean. For example, suppose that the Atlanta consumer group wishes to express its estimate as an interval at the .95 confidence level. A confidence level of .95 implies that there is a .95 probability that the true population mean lies within 1.96 standard deviations (an area of .4750 on either side) of the sample mean. The limits of this interval can be found on the sampling distribution of the means of samples of size 30, as shown in Figure 6-1.

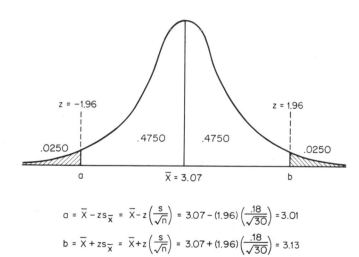

$$a = \overline{X} - zs_{\overline{X}} = \overline{X} - z\left(\frac{s}{\sqrt{n}}\right) = 3.07 - (1.96)\left(\frac{.18}{\sqrt{30}}\right) = 3.01$$

$$b = \overline{X} + zs_{\overline{X}} = \overline{X} + z\left(\frac{s}{\sqrt{n}}\right) = 3.07 + (1.96)\left(\frac{.18}{\sqrt{30}}\right) = 3.13$$

Figure 6-1   Interval estimate of a population mean, normal distribution.

Strictly speaking, the sampling distribution has a mean of μ, and $\overline{X}$ is the random variable. In interval estimation, it is more convenient to think of μ as a variable distributed about $\overline{X}$. Thus, we find the limits of the interval by

$$\overline{X} \pm z\sigma_{\overline{X}} \tag{6-1}$$

if σ is known. If σ is not known, it must be estimated from s, and the standard error of the mean is estimated by $s/\sqrt{n}$ and designated $s_{\overline{X}}$. The interval limits in this case are

$$\overline{X} \pm zs_{\overline{X}} \tag{6-2}$$

Additionally, if σ is estimated and the sample is small, the t distribution applies and the interval is given by

$$\overline{X} \pm ts_{\overline{X}} \tag{6-3}$$

In the coffee example, the large samples suggest a normal sampling distribution and the use of either formula (6-1) or (6-2). The choice is resolved by the unknown population standard deviation which requires an estimated standard error of the mean $s_{\overline{X}}$, as shown in formula (6-2).

The .95 interval estimate of the mean price of a 1-pound can of coffee in Atlanta then is \$3.01 to \$3.13. Another interpretation of this estimate is that there is only a 5 percent chance $(1 - .95)$ that the true population mean is outside the interval, that is, less than \$3.01 or greater than \$3.13. This probability of error is represented by the shaded tails in Figure 6-1 and is identified with the lowercase Greek letter α (alpha). Alpha includes *both* tails; a single tail is one-half of alpha, or $\alpha/2$. The designation of error probabilities becomes more important later in hypothesis testing.

Suppose the sample of coffee prices consists only of the first 10 observations in Table 6-2. Now, the sample mean and sample standard deviation are

$$\overline{X} = \frac{\Sigma X}{n} = \frac{30.44}{10} = 3.044$$

$$s = \sqrt{\frac{\Sigma X^2 - (\Sigma X)^2/n}{n-1}} = \sqrt{\frac{92.8088 - 92.6594}{9}} = .12886$$

The sampling distribution of these small samples with an unknown population standard deviation is no longer normal; it is described by the t distribution, provided the population is normal, an assumption we now make.

The appropriate formula for an interval estimate based on a t distribution is (6-3). In addition to the sample mean, which has already been computed, this equation requires two additional terms

$$s_{\bar{X}} = \frac{s}{\sqrt{n}} = \frac{.12886}{\sqrt{10}} = .041 \quad \text{and} \quad t_{(.025,9)} = 2.262$$

The subscripted t follows the general form $t_{(\alpha/2,df)}$, in which $\alpha/2$ is the area of one tail and df represents the degrees of freedom. These are, of course, the two arguments for entering Appendix G where $t_{(.025,9)}$ is found as 2.262.

The new .95 interval estimate of the population mean coffee price, shown in Figure 6-2, is given by

$$\bar{X} \pm t s_{\bar{X}}$$
$$3.044 \pm (2.262)(.041)$$
$$2.951 \text{ to } 3.137$$

or approximately 2.95 to 3.14

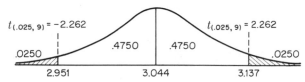

Figure 6-2   Interval estimate of a population mean, t distribution.

In some cases, particularly in production control applications, it is not unusual to know the population standard deviation but not the population mean. This frequently occurs in a mechanical process where the mean varies as the process gets out of tolerance but the distribution of errors about the mean is constant. It is somewhat analogous to a rifle that has a constant shot pattern, the center of which moves about as the rifle is aimed at different points. To illustrate an interval estimate of such a process, consider a batch of steel plates rolled out by a machine that tends to give a distribution of plate thickness with a standard deviation of .025 inch. Furthermore, suppose that there are 1500 plates in the batch and a sample of 100 gives a mean thickness of .125 inch. What is the .90 interval estimate of the mean thickness of all 1500 plates?

The large sample size causes the sampling distribution to be normal and the known standard deviation permits an exact rather than an estimated computation of the standard error of the mean, suggesting the use of formula (6-1). However, n/N is $100/1500 = .067$, a value larger than the nominal .05 and one which implies use of the finite-population correction. The standard error of the mean is therefore

$$\sigma_{\bar{X}} = \frac{\sigma}{\sqrt{n}} \sqrt{\frac{N-n}{N-1}}$$
$$= \frac{.025}{\sqrt{100}} \sqrt{\frac{1500-100}{1500-1}}$$
$$= .0024$$

Equation (6-1) also requires a z value—in this case, one based on a 90 percent interval centered on the mean, as shown in Figure 6-3. This gives z values of ± 1.64, and the interval can be computed:

$$\bar{X} \pm z\sigma_{\bar{X}}$$
$$.125 \pm (1.64)(.0024)$$
$$.121 \text{ to } .129$$

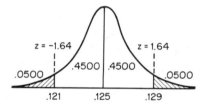

Figure 6-3   Interval estimate of a population mean, finite population.

That is, there is a 90 percent probability that the population mean thickness is between .121 and .129 inch.

The rules for finding an interval estimate under various combinations of sample size and population characteristics are summarized later in Table 6-3.

The Determination of Sample Size

As the confidence level in an interval estimate increases, the interval itself becomes larger. While the confidence that results from a large interval may be gratifying, too large an interval can make the estimate almost useless for decision-making purposes. For example, a 99 percent confidence interval for the estimated volume of automobile sales in the United States next year could represent a recession at the lower limit and a boom at the upper limit. Reducing the confidence level of the estimate will narrow the range but will add uncertainty. How can a manager increase the accuracy of an estimate and still maintain a desired level of confidence? The answer is simple: increase the sample size.

As sample size increases, the standard error of the mean decreases, and variability in the sampling distribution is reduced. This implies that a given number of standard errors will result in upper and lower limits closer to the mean and the interval will be smaller. This is especially true if the sample-size increase results in a change from the t to the normal distribution, which is more concentrated about the mean. To illustrate this point, consider the 95 percent confidence interval about a sample mean of 100 when the sample standard deviation is 12 and the sample size is n:

$n = 4$:

$$\hat{\mu} = \bar{X} \pm t_{(.025,3)} \frac{s}{\sqrt{n}}$$

$$= 100 \pm (3.182) \frac{12}{\sqrt{4}} = 100 \pm 19.092$$

$$= 80.908 \text{ to } 119.092$$

$n = 16$:

$$\hat{\mu} = \bar{X} \pm t_{(.025,15)} \frac{s}{\sqrt{n}}$$

$$= 100 \pm (2.131) \frac{12}{\sqrt{16}} = 100 \pm 6.393$$

$$= 93.607 \text{ to } 106.393$$

$n = 36$:

$$\hat{\mu} = \bar{X} \pm z \frac{s}{\sqrt{n}}$$

$$= 100 \pm (1.96) \frac{12}{\sqrt{36}} = 100 \pm 3.92$$

$$= 96.08 \text{ to } 103.92$$

$n = 100$:

$$\hat{\mu} = \bar{X} \pm z \frac{s}{\sqrt{n}}$$

$$= 100 \pm (1.96) \frac{12}{\sqrt{100}} = 100 \pm 2.352$$

$$= 97.648 \text{ to } 102.352$$

Increasing sample size from 4 to 16 to 36 to 100 decreased the interval[2] from approximately 38 to 13 to 8 to 5. Clearly, the most dramatic

---

[2]The interval is twice $t s_{\bar{x}}$ or $z s_{\bar{x}}$. For $n = 4$, the interval is $2 \times 19.092 = 38.184$, or approximately 38. Other intervals can be approximated similarly.

decreases in the interval occur early. Since sampling can be expensive, a manager may not wish to increase the sample size from 36 to 100 just to decrease the interval from 8 to 5.

It is also possible to compute the sample size required to maintain a desired interval. For example, suppose that an accountant wants an estimate of the mean balance of accounts receivable. Previous experience has shown that the mean varies but the standard deviation is relatively constant at $120. How many accounts must be sampled to keep the interval to $20 or less at the 90 percent confidence level?

Since an interval estimate is always symmetrical about the sample mean, an interval of $20 suggests $10 on either side of the mean; that is, the upper limit is $\overline{X} + \$10$ and the lower limit is $\overline{X} - \$10$. We also know that the upper limit can be expressed as $\overline{X} + z\sigma_{\overline{x}}$, or $\overline{X} + (1.64)(120/\sqrt{n})$. The two expressions for the upper limit can be set equal and solved for n:

$$\overline{X} + 10 = \overline{X} + (1.64)\frac{120}{\sqrt{n}}$$

$$10 = (1.64)\frac{120}{\sqrt{n}} \qquad \text{(Subtract } \overline{X} \text{ from both sides)}$$

$$10\sqrt{n} = (1.64)(120) \qquad \text{(Multiply both sides by } \sqrt{n})$$

$$\sqrt{n} = \frac{(1.64)(120)}{10} \qquad \text{(Divide both sides by 10)}$$

$$\sqrt{n} = 19.68 \qquad \text{(Multiply, divide)}$$

$$n = 387.3 \text{ or } 388 \qquad \text{(Square both sides)}$$

Notice that to ensure that the interval is $20 *or less* n must be rounded up to the next higher whole number. To· round down, however slightly, would make the interval somewhat larger than $20.

The half-interval distance from the mean to the upper and lower limits is often treated as a maximum allowable error E. By substituting E for the $10 in the previous example, a generalized formula for n can be developed:

$$\overline{X} + E = \overline{X} + z\frac{\sigma}{\sqrt{n}}$$

$$E = z\frac{\sigma}{\sqrt{n}}$$

$$E\sqrt{n} = z\sigma \qquad\qquad (6\text{-}4)$$

$$\sqrt{n} = \frac{z\sigma}{E}$$

$$n = \left(\frac{z\sigma}{E}\right)^2$$

To illustrate the use of this formula, assume that the accountant in the previous example will accept a $15 error and a 95 percent confidence interval ($z = 1.96$). The minimum sample required to achieve an estimate within $15 is

$$
\begin{aligned}
n &= \left(\frac{z\sigma}{E} \right)^2 \\
&= \left[\frac{(1.96)(120)}{15} \right]^2 \\
&= (15.68)^2 \\
&= 245.86 \text{ or } 246
\end{aligned}
$$

HYPOTHESIS TESTING

The concept of hypothesis testing is rooted in scientific research, where, classically, the results of an experiment are compared with a hypothetical outcome and the validity of the hypothesis is assessed accordingly. Most managerial decisions take place in environments that cannot be controlled as rigidly as the laboratory, but the general framework of hypothesis testing is still a valuable managerial tool. If a population possesses certain characteristics (parameters), one decision may be indicated; if those characteristics do not exist, another decision may be more appropriate.

It is but a short transitional step from interval estimation to hypothesis testing. In interval estimation, an interval is constructed about a *sample mean* to give the limits of the estimated population mean. In hypothesis testing, a similar interval is constructed about a *hypothesized mean* to see whether a sample mean is or is not included. In keeping with its scientific heritage, a hypothesis test is usually conducted in five strictly defined steps:

1. Formulate the hypothesis

2. Establish a significance level for testing

3. Formulate a decision rule

4. Take a sample

5. Make a decision

A Two-Tail Hypothesis Test

To illustrate these five steps, let us suppose that a manufacturer produces flashbulbs in lots of 50,000. Although the mean intensity of the flash varies from lot to lot, the standard deviation tends to be a constant 200 lumens. The flashbulbs are sold for use in a camera that has settings calibrated for a flash intensity of 5000 lumens. Before being released for distribution, each lot is tested for conformance to the design specifications. Since testing is destructive, the manufacturer will test only 100 flashbulbs from each lot. He is willing to risk a 5 percent chance of rejecting a lot when the mean intensity actually is 5000 lumens.

Formulate the Hypothesis   The hypothesis to be tested is called the *null hypothesis* and is identified by H_0. The term "null" is interpreted as a no-change condition; that is, it is an expression of what one expects under normal conditions.[3] In this case, no change from normal conditions means that the population mean is 5000 lumens and we write the null hypothesis as

$$H_0: \mu = 5000$$

Establish a Significance Level   The *significance level* of a hypothesis test is the probability that the null hypothesis will be rejected when it is true, an error condition we address in greater detail later in this chapter. The significance level must be established subjectively by the decision maker. In this case, the manufacturer was actually specifying the significance level when he accepted a 5 percent chance of rejecting a lot with the correct mean of 5000 lumens. The significance level corresponds to the tail area of a sampling distribution and is identified with the Greek letter α, as in interval estimation:

$$\alpha = .05$$

Because a sample mean that is either too high or too low will cause the null hypothesis to be rejected, the significance level must be split between the upper and lower tails of the sampling distribution. This so-called *two-tail* hypothesis test points up another similarity to interval estimation, a symmetrical interval about the mean. It also suggests a relationship between the confidence level and the significance level, namely, *confidence level + significance level = 1* or, as it is more commonly expressed, *1 − confidence level = significance level*.

Formulate a Decision Rule   For a two-tail test, the null hypothesis is rejected when the sample mean falls beyond either of two *critical values*

---

[3]There are various approaches to the formulation of hypotheses, some of which require a second or *alternate* hypothesis (H_1 or H_a) to convey the complete meaning of the null hypothesis. In the approach described here there is no need for an alternate hypothesis, and none will be used. If we did include an alternate hypothesis, it would be collectively exhaustive with the null hypothesis; that is, it would cover all situations not included in the null hypothesis. Thus, for $H_0: \mu = 5000$, $H_a: \mu \neq 5000$.

that correspond to the upper and lower limits of a $1 - \alpha$ interval about the hypothesized mean.

When the sampling distribution is normal, the 5 percent significance level results in z values of ± 1.96, the same values associated with a 95 percent confidence level. The corresponding critical values for this problem now become

$$CV - \pm z\sigma_{\overline{X}}$$
$$= 5000 \pm (1.96)\frac{200}{\sqrt{100}}$$
$$= 5000 \pm (1.96)(20)$$
$$= 4960.8 \text{ and } 5039.2$$

The areas of the sampling distribution to the left of 4960.8 lumens and to the right of 5039.2 lumens are called *rejection regions,* because, for this problem, a sample mean in either of these two areas will result in rejection of the null hypothesis. Strictly speaking, a sample mean that falls between the critical values does not result in acceptance of the null hypothesis but causes the null hypothesis not to be rejected. The rationale for this seemingly minor semantic point is that a sample mean of, say, 4085 lumens certainly does not prove that the population mean is 5000 lumens. But 4085 is within a 95 percent confidence interval about 5000, and the null hypothesis cannot be rejected at the .05 significance level.

The difference between accepting the null hypothesis and not rejecting it is somewhat academic to the flashbulb manufacturer. His decision cannot be qualified. He must either accept the lot or reject it. In this case, not rejecting the lot is tantamount to accepting it, and accordingly we identify the area between critical values as the *acceptance region.* The decision rule for this test can now be stated:

Test a sample of 100 flashbulbs. If the mean intensity is between 4960.8 and 5039.2 lumens, accept the lot; otherwise, reject it.

Take a Sample   From the lot of flashbulbs, the quality control manager of the firm randomly selects 100 flashbulbs and measures their light intensity. Let us assume that the mean intensity of the sample is

$$\overline{X} = 4980$$

Make a Decision   Figure 6-4 shows that the sample mean of 4980 lumens falls in the acceptance region. The null hypothesis should not be rejected, and the lot should be accepted. Of course, if the sample mean were in the rejection region, the decision would be unambiguous: both the null hypothesis and the lot would be rejected.

A One-Tail Hypothesis Test

The test of the hypothesis concerning the mean intensity of flashbulbs involved two tails or rejection regions because a mean was unacceptable if

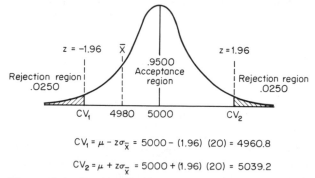

$$CV_1 = \mu - z\sigma_{\bar{x}} = 5000 - (1.96)\ (20) = 4960.8$$

$$CV_2 = \mu + z\sigma_{\bar{x}} = 5000 + (1.96)\ (20) = 5039.2$$

Figure 6-4   Two-tail hypothesis test.

it was either too high or too low, a situation suggested by the equals sign in the null hypothesis. In some cases, rejection is appropriate only when the mean is too high (mean tar content of cigarettes) or too low (mean life of light bulbs). For tests of hypotheses concerning these means, there will be only one tail or rejection region (hence the term *one-tail hypothesis test*), and the null hypothesis will be expressed as an *inequality.*

To illustrate a one-tail hypothesis test, consider a cereal producer who wishes to test the hypothesis that her cereal has a mean of at least 250 raisins per box. An indication that the mean is less than 250 requires a shutdown of the packaging process to recalibrate the raisin dispenser, a situation she is willing to accept only 1 percent of the time when the true mean is 250 or more. The standard deviation of the population is unknown and must be estimated from the sample of 64 boxes.

Formulate the Hypothesis   The key words in this problem are "a mean of at least 250 raisins." *At least* implies 250 or more in the null condition. The null hypothesis is therefore

$$H_0: \mu \geqslant 250$$

Establish a Significance Level   The producer has specified a 1 percent chance of shutting down the packaging process when it is in fact operating properly. This corresponds to the significance level definition of rejecting the null hypothesis when it is true. Therefore,

$$\alpha = .01$$

Formulate a Decision Rule The inequality expressed in the null hypothesis indicates a one-tail test, a single rejection region, and only one critical value. The inequality also indicates the general location of that critical value: a *less-than-or-equal-to* null hypothesis implies a critical value to the *right* of the mean while a *greater-than-or-equal-to* null hypothesis implies one to the *left*. The rationale for this argument is apparent in the raisin cereal case. Obviously no level of raisin content *above* 250 could ever be construed as an indication that the true mean is not 250 or greater, and because of sampling error some values *below* 250 must be accepted also. The critical value must therefore be *less* than 250. Similar logic can be applied to the case for establishing a critical value above the mean when the null hypothesis contains a less-than-or-equal-to expression.

Even though this test involves only one tail, the area of the rejection region still corresponds to the significance level of .01. The only difference is that the one tail contains all the rejection region whereas each tail in the two-tail test contained one-half of the total rejection region. From Table 4-8, we see that the z value for a 1 percent left-hand tail (99 percent greater than negative z) is -2.33.

Since the population standard deviation is unknown and sample data are not yet available to estimate it, we temporarily state the decision rule in terms of a z value rather than a number of raisins:

Count the raisins in a sample of 64 boxes. If the sample mean is more than 2.33 standard errors of the mean below 250 raisins, stop the packaging process and recalibrate the raisin dispenser; otherwise, let it continue.

Take a Sample Assume that the sample of 64 boxes gives a mean raisin count of 240 with a sample standard deviation of 24. The single critical

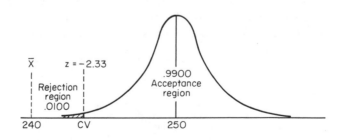

$$CV = \mu - zs_{\bar{x}} = 250 - (2.33)(3.00) = 243$$

Figure 6-5 One-tail hypothesis test.

value can now be computed, as shown in Figure 6-5, using the estimated standard error of the mean

$$CV = \mu - zs_{\bar{X}}$$
$$= 250 - (2.33)\frac{24}{\sqrt{64}}$$
$$= 250 - (2.33)(3.00)$$
$$= 242$$

Make a Decision   The sample mean is less than the critical value and the null hypothesis, as well as this batch of raisin cereal, is rejected.[4]

Hypothesis Tests with Small Samples or Finite Populations

In the preceding examples, the samples were large enough to permit the use of the normal distribution, and the populations were large enough to make the finite-sample correction factor unnecessary. Since hypothesis testing is so closely related to interval estimation, it is not surprising to find that the various sampling distribution rules (Table 5-8) that governed estimation also apply to hypothesis testing. Thus, when $n < 30$, σ is unknown, and the population is normal, the t distribution is used instead of the normal. And when $n/N \geq .05$, the finite-sample correction factor is applied to both the standard error of the mean $\sigma_{\bar{X}}$ and its estimator $s_{\bar{X}}$.

When sample size and the sample-to-population ratio are considered along with one- and two-tail tests and known and unknown population standard deviations, there is a great variety of combinations for the decision maker to consider. The computation of the critical values for these various combinations is shown in Table 6-3 and three additional examples, illustrated in Figure 6-6, are explained below.

An automobile rental firm wishes to test the hypothesis, at the .05 significance level, that the mean annual maintenance cost for 700 cars in its fleet is $175 or less. A sample of 25 cars gives a mean of $187 and a standard deviation of $52. Should the hypothesis be accepted?

σ is unknown, n is small, the population is assumed to be normal, and $n/N = 25/700 = .036$.

$$H_0: \mu \leq 175 \text{ (one-tail test)}$$
$$s_{\bar{X}} = \frac{52}{\sqrt{25}} = 10.40$$
$$t_{(.05,24)} = 1.711$$
$$CV = \mu + ts_{\bar{X}} = 175 + (1.711)(10.40) = 192.80$$

[4]An alternate method for applying the decision rule in cases where the critical value is expressed as a z value is to express the sample mean as a z value as well. In this case, the sample $z = (\bar{X} - \mu)/s_{\bar{X}}$, or $(240 - 250)/3 = -3.33$. Since -3.33 lies in the rejection region (to the left of -2.33), H_0 is rejected.

TABLE 6-3   Summary of equations for interval estimation and hypothesis testing*

n	Population	σ	Interval estimation Interval	Hypothesis testing $H_0: \mu \leq X$	$H_0: \mu \geq X$	$H_0: \mu \geq X$
≥ 30	Any	Known	$\overline{X} \pm z\sigma_{\overline{X}}$	$\mu \pm z\sigma_{\overline{X}}$	$\mu + z\sigma_{\overline{X}}$	$\mu - z\sigma_{\overline{X}}$
		Unknown	$\overline{X} \pm zs_{\overline{X}}$	$\mu \pm zs_{\overline{X}}$	$\mu + zs_{\overline{X}}$	$\mu - zs_{\overline{X}}$
<30	Normal	Known	$\overline{X} \pm z\sigma_{\overline{X}}$	$\mu \pm z\sigma_{\overline{X}}$	$\mu + z\sigma_{\overline{X}}$	$\mu - z\sigma_{\overline{X}}$
		Unknown	$\overline{X} \pm ts_{\overline{X}}$	$\mu \pm ts_{\overline{X}}$	$\mu + ts_{\overline{X}}$	$\mu - ts_{\overline{X}}$

*Both $\sigma_{\overline{X}}$ and $s_{\overline{X}}$ include the finite-population correction $\sqrt{(N - n)/(N - 1)}$ when $n/N \geq .05$.

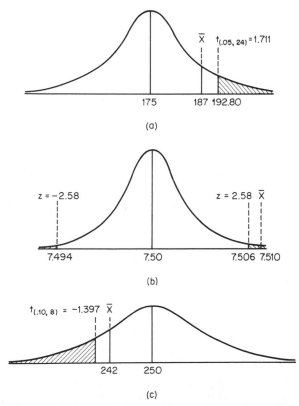

(a)

(b)

(c)

Figure 6-6   Hypothesis tests for small samples and/or finite populations.

Since $\overline{X} = \$187$ is less than the critical value of $\$192.80$, do not reject the null hypothesis (Figure 6-6a).

A sample of 81 from an order of 1000 two-by-four studs used for framing buildings has a mean length of 7 feet, 6⅛ inches (7.510 feet). Previous experience suggests that the standard deviation of the cutting process is a constant ¼ inch (.021 foot). At the .01 significance level, should the hypothesis that the mean of the order is 7 feet 6 inches (7.500 feet) be accepted?

σ is known, n is large, and $n/N = 81/1000 = .081$.

$H_0:\ \mu = 7.500$ (two-tail test)

$$\sigma_{\overline{X}} = \frac{\sigma}{\sqrt{n}} \sqrt{\frac{N-n}{N-1}} = \frac{.021}{\sqrt{81}} \sqrt{\frac{1000-81}{1000-1}} = (.0023)(.96) = .0022$$

$z = \pm 2.58$

$\text{CV} = \mu \pm z\sigma_X = 7.500 \pm (2.58)(.0022) = 7.494$ and 7.506

Since $\overline{X} = 7.510$ feet is greater than the critical value of 7.506 feet, the null hypothesis should be rejected (Figure 6-6b).

A contract for 150 space suits for astronauts specifies that they must withstand a mean pressure of at least 250 pounds per square inch (psi). Since the suits are very expensive, only nine are tested by inflating them until they burst. If the sample mean bursting point is 242 psi with a standard deviation of 21 psi, should the order be accepted at the .10 significance level?

σ is unknown, n is small, the population is assumed to be normal, and $n/N = 9/150 = .06$.

$H_0:\ \mu \geqslant 250$ (one-tail test)

$$s_{\overline{X}} = \frac{s}{\sqrt{n}} \sqrt{\frac{N-n}{N-1}} = \frac{21}{\sqrt{9}} \sqrt{\frac{150-9}{150-1}} = (7)(.97) = 6.79$$

$t_{(.10,8)} = -1.397$

$\text{CV} = \mu - ts_{\overline{X}} = 250 - (1.397)(6.79) = 240.5$

Since $\overline{X} = 242$ psi is greater than the critical value of 240.5 psi, the null hypothesis cannot be rejected (Figure 6-6c). Whether or not the space suits should be accepted is questionable. The manufacturer would certainly claim that the hypothesis test argues for acceptance. The customer,

NASA, might take a different point of view. Suppose that NASA conducted the same test with the same sample results, but used the null hypothesis

$$H_0: \mu \leqslant 250$$

Now the critical value is *above* the mean

$$\text{CV} = \mu + ts_{\bar{x}} = 250 + (1.397)(6.79) = 259.5$$

and both the hypothesis and the space suits would be rejected. Clearly, the formulation of the null hypothesis can depend on one's point of view and the decision rule must be defined precisely in any contract that depends on an acceptance test.

Errors in Hypothesis Testing

The error of rejecting the null hypothesis when it is true, a possibility addressed in each of the preceding examples, is called a *Type I error*. It is also an error to not reject the null hypothesis when it is false, a mistake described as a *Type II error*. The conditions that lead to Type I and Type II errors are summarized in Table 6-4.

TABLE 6-4 Error conditions in hypothesis testing

Decision	H_0 is true	H_0 is false
Reject H_0	Type I error	Correct decision
Do not reject H_0	Correct decision	Type II error

Type I Errors The maximum probability of a Type I error is α, the significance level. This probability is commonly called the *producer's risk*, since a producer testing his own product may discard an acceptable lot by committing a Type I error.

When the null hypothesis involves an equality, as in the flashbulb example, the probability of a Type I error is *exactly* α, the combined tail areas of a two-tail test. However, when the null hypothesis involves an inequality, the tail area, and consequently the value of α, varies with the true population mean. For example, a .10 significance level test of the hypothesis that the mean life of a very large shipment of batteries is equal to or greater than 1000 hours, based on a population standard deviation of 200 hours and a sample of 100 batteries, would have a critical value of 974.4 hours. When the population mean is *exactly* 1000 hours, the probability of rejecting the shipment is *exactly* 10 percent, as depicted by the uppermost distribution in Figure 6-7.

The subsequent distributions in Figure 6-7 show that the tail area α, the probability of a Type I error, decreases as the true population mean increases to 1010, 1020, and 1030 hours. The four means and their corresponding α values have been plotted in Figure 6-8. The resultant curve, called a *Type I error curve*, can be used to find the value of α for any true population mean between 1000 hours and the point where α approaches zero, approximately 1040 hours.[5]

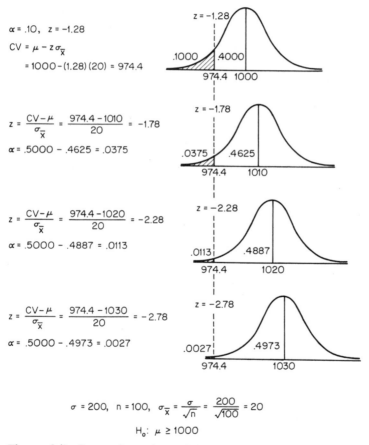

$$\alpha = .10, \quad z = -1.28$$

$$CV = \mu - z\sigma_{\bar{x}}$$

$$= 1000 - (1.28)(20) = 974.4$$

$$z = \frac{CV - \mu}{\sigma_{\bar{x}}} = \frac{974.4 - 1010}{20} = -1.78$$

$$\alpha = .5000 - .4625 = .0375$$

$$z = \frac{CV - \mu}{\sigma_{\bar{x}}} = \frac{974.4 - 1020}{20} = -2.28$$

$$\alpha = .5000 - .4887 = .0113$$

$$z = \frac{CV - \mu}{\sigma_{\bar{x}}} = \frac{974.4 - 1030}{20} = -2.78$$

$$\alpha = .5000 - .4973 = .0027$$

$$\sigma = 200, \quad n = 100, \quad \sigma_{\bar{x}} = \frac{\sigma}{\sqrt{n}} = \frac{200}{\sqrt{100}} = 20$$

$$H_0: \mu \geq 1000$$

Figure 6-7   Decreasing values of the producer's risk for increasing values of the true population mean.

[5]It is also possible to compute the probability of rejecting the null hypothesis when it is false, a correct decision. When these probabilities and population means are added to the Type I error curve, it is called a *power curve*.

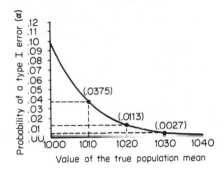

Figure 6-8 Type I error curve.

Type II Errors Just as some sample means will fall in the rejection region when the null hypothesis is true, many sample means will fall in the acceptance region when the null hypothesis is not true. In the latter case, the null hypothesis will be accepted (or, at least, not rejected), an error of Type II. The probability of such an error is called the *consumer's risk* because a producer accepting a product as having a hypothesized mean when it actually does not may pass a defective product along to the consumer.

The consumer's risk is identified by the Greek letter β (beta) and, like α, its exact value is dependent upon the value of the true population mean. Since the null hypothesis must be false for a Type II error to occur, we shall consider only population means other than those hypothesized. For the example of mean battery life, only population means less than 1000 hours can result in a Type II error. The probability that the null hypothesis will be accepted is equal to the area of the sampling distribution that extends into the acceptance region. These probabilities are shown for true population means of 980, 960, 940, and 920 hours in Figure 6-9.

An error curve for the probability of a Type II error at various true population means can now be constructed.[6] Such a curve has been combined with the Type I error curve in Figure 6-10 to show the probability of *either* type of error for the full range of possible true population means.

Sample Size in Hypothesis Testing

The determination of an adequate sample size for an interval estimate was based upon a maximum allowable error. A similar procedure is followed in selecting the size of a sample to test a hypothesis. The principal differences between the two methods lie in the number and form of the errors; there is one error in interval estimation, usually expressed in the

[6]Probabilities of accepting the null hypothesis when it is true, a correct decision, can be added to the Type II error curve to make what is called an *operating-characteristics* curve.

same units as the sample mean, while hypothesis testing involves two errors expressed as probabilities.

To demonstrate the process of sample-size selection with the battery-life example, suppose that the quality control manager is willing to reject the shipment 5 percent of the time when the true mean is 1000 hours (a Type I error) and to accept it 15 percent of the time when the true mean is only 950 hours (a Type II error).[7] By using a population standard devia-

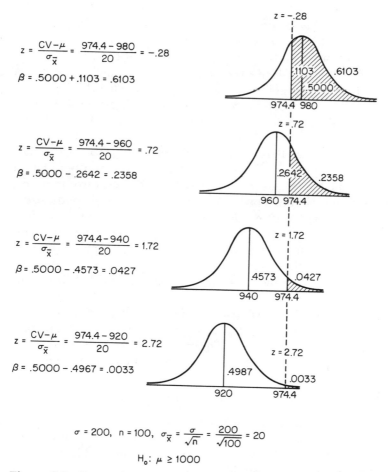

$$z = \frac{CV - \mu}{\sigma_{\bar{x}}} = \frac{974.4 - 980}{20} = -.28$$

$$\beta = .5000 + .1103 = .6103$$

$$z = \frac{CV - \mu}{\sigma_{\bar{x}}} = \frac{974.4 - 960}{20} = .72$$

$$\beta = .5000 - .2642 = .2358$$

$$z = \frac{CV - \mu}{\sigma_{\bar{x}}} = \frac{974.4 - 940}{20} = 1.72$$

$$\beta = .5000 - .4573 = .0427$$

$$z = \frac{CV - \mu}{\sigma_{\bar{x}}} = \frac{974.4 - 920}{20} = 2.72$$

$$\beta = .5000 - .4967 = .0033$$

$$\sigma = 200, \quad n = 100, \quad \sigma_{\bar{x}} = \frac{\sigma}{\sqrt{n}} = \frac{200}{\sqrt{100}} = 20$$

$$H_o: \mu \geq 1000$$

Figure 6-9 Decreasing values of the consumer's risk for decreasing values of the true population mean.

[7]This problem is presented from the point of view of the *producer's* quality control manager. The quality control manager for the *consumer* most likely would reverse the probabilities.

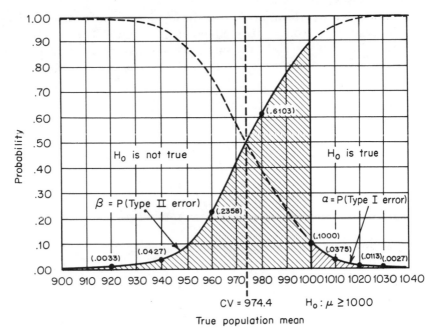

Figure 6-10 Type I and Type II error curves.

tion of 200 hours the following equations can be written for the critical value to be used in this test:

Type I:

$$\mu = 1000 \qquad \alpha = .05 \qquad z = -1.64 \qquad (\textit{One-tail} \text{ test})$$

$$z = \frac{\text{CV} - \mu}{\sigma_{\bar{x}}}$$

$$-1.64 = \frac{\text{CV} - 1000}{200/\sqrt{n}}$$

$$\text{CV} = 1000 - \frac{(1.64)(200)}{\sqrt{n}}$$

Type II:

$$\mu = 975 \qquad \beta = .15 \qquad z = 1.04 \qquad (\text{Appendix F})$$

$$z = \frac{\text{CV} - \mu}{\sigma_{\bar{x}}}$$

$$1.04 = \frac{\text{CV} - 950}{200/\sqrt{n}}$$

$$\text{CV} = 950 + \frac{(1.04)(200)}{\sqrt{n}}$$

In their present form, neither of these equations can be solved separately because each contains two unknowns, CV and n. They can be solved simultaneously, however, by the simple expedient of subtracting one from the other to eliminate the CV term:

Type I:
$$CV = 1000 - \frac{(1.64)(200)}{\sqrt{n}}$$

Type II:
$$- \quad CV = 950 + \frac{(1.04)(200)}{\sqrt{n}}$$

$$\overline{\qquad\qquad\qquad\qquad\qquad\qquad}$$

$$0 = 50 - \frac{2.68(200)}{\sqrt{n}}$$

Now this single equation can be solved for n:

$$\frac{(2.68)(200)}{\sqrt{n}} = 50$$

$$\sqrt{n} = \frac{(2.68)(200)}{50} = 10.72$$

$$n = 114.92 \approx 115$$

and n can be substituted into either of the original equations to solve for the critical value:

Type I:
$$CV = 1000 - \frac{(1.64)(200)}{\sqrt{115}}$$

$$= 1000 - \frac{(1.64)(200)}{10.72} = 969.40$$

A pictorial representation of this test is shown in Figure 6-11.

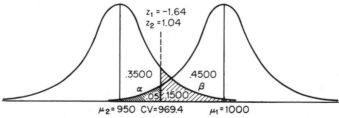

Figure 6-11   Hypothesis test at $\alpha = .05$, $\beta = .15$, and $n = 460$.

The formula for a hypothesis-test sample size can be developed symbolically to avoid the algebra of solving simultaneous equations for each hypothesis test. The generalized formula takes the form

$$n = \left[\frac{(z_2 - z_1)\sigma}{\mu_1 - \mu_2} \right]^2 \tag{6-5}$$

where $z_1 =$ z value for probability α of Type I error

$z_2 =$ z value for probability β of Type II error

$\mu_1 =$ population mean associated with α

$\mu_2 =$ population mean associated with β

For the battery-life test, n could have been computed using this formula

$$n = \left[\frac{(1.04 - (-)1.64)(200)}{1000 - 950} \right]^2$$

$$= \left[\frac{(2.68)(200)}{50} \right]^2 = 114.92 \approx 115$$

Other Hypothesis Tests

In this chapter we have treated only tests of hypotheses concerning the mean of a population of continuous data. The principles involved are also applicable to tests of other parameters and, with slight modification, to populations of discrete data. We also test hypotheses concerning the parameters of a linear-regression analysis in Chapter 7. Thus, as we have noted earlier, additional topics in statistics typically build on a few essentials. Hypothesis testing is one essential which, along with estimation and probability, has repeated applications to managerial decision making.

CHAPTER 7

Multivariate Analysis: Linear Regression and Correlation Analysis

The statistical analyses described in the preceding four chapters are based on *univariate* data, that is, data consisting of only one characteristic or variable per observation. In various examples, we considered the length of employment of workers in a factory, the age of employees in a small business, the number of fares for taxis, and so on. But these characteristics were always considered *singly*; we did not, for example, consider both the age and length of employment of workers in the same analysis.

When there are two or more variables per observation, the data are described as *multivariate*. Multivariate analyses can become cumbersome both from their conceptual complexity and the sheer bulk of computational processes normally associated with them. For both reasons, serious analyses of multivariate data are usually performed with the aid of a computer. The purpose of this chapter is to give the practical background required to interpret and apply the results of a computer-generated analysis of multivariate data but not necessarily to teach the skills required to perform that analysis manually.

UNIVARIATE ANALYSIS

Let us approach the problem of multivariate analysis as a logical extension of some univariate analyses described in Chapters 3 and 6. Suppose the sales manager of a medium-sized life insurance firm has collected data on the amount of insurance sold last year by a sample of 10 agents. The amount sold by each agent is shown in Table 7-1.

If another agent is hired to operate in a new region, what additional sales would be expected? Intuitively, we suspect that life insurance sales are related to the experience, skill, and aggressiveness of the agent; the age, occupation, and economic status of prospective clients; the level of activity of competitive firms; and many other factors. But without this information, we can only base an estimate on the data at hand, the sales of the 10 agents in the sample. Of the several measures of central tendency on which an estimate could be based, we choose the mean (there is no mode, there are no extreme values to suggest use of the median, and we anticipate further statistical computations that require the mean):

$$\hat{Y} = \bar{Y} = \frac{\Sigma Y}{n} = \frac{12.1}{10} = 1.21$$

TABLE 7-1   Life insurance sales in millions of dollars for a sample of 10 sales agents

Agent	Sales Y	Agent	Sales Y
1	.62	6	1.19
2	.81	7	1.31
3	.94	8	1.45
4	1.03	9	1.68
5	1.15	10	1.92

If no other information were available, we would predict sales of $1.21 million for the new agent. If population data were available and the distribution was normal, we could construct a confidence interval about this prediction. For example, assume the population mean μ is the same as the sample mean of $1.21 million and the population standard deviation σ_Y is the same as the sample standard deviation s_Y which, in this case, is approximately $.396 million.[1] A 90 percent confidence interval about the prediction would be

$$\hat{Y} = \mu \pm z\sigma_Y$$
$$= 1.21 \pm (1.64)(.396)$$
$$= .560560 \text{ to } 1.859440$$

or $560,560 to $1,859,440.

The sales data are shown graphically in Figure 7-1, along with the point and interval estimates. Notice that univariate data can be represented in

---

[1]The symbol s_Y is used here to designate the standard deviation of a sample of values of Y. The use of of subscripts to distinguish between similar symbols is common in the regression analysis which follows.

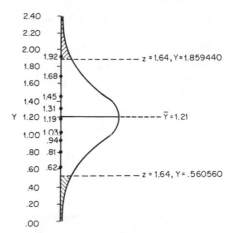

Figure 7-1 Point and 90 percent confidence interval estimates of sales based on $\overline{Y}$, s_Y.

a single dimension, in this case, a vertical line that corresponds to the Y axis of a graph. The designation of sales as Y instead of the more common random variable X is in keeping with graphing conventions that will be explained later.

Figure 7-1 illustrates an inherent weakness in this univariate analysis. With all the variance of Y free to influence the estimate, the interval is almost as great as the range of the sample data! One of the major objectives in subsequent analysis is to reduce variance by introducing explanatory variables.

MULTIVARIATE ANALYSIS

The sales manager considers the interval estimate to be too vague and wishes to reduce the interval without changing the confidence level or sample size. Upon examining personnel records and demographic studies, the sales manager finds additional information about the agents and their sales regions that may explain some of the variance in sales.

The additional information is more than just a few more data sets, each to be subjected to a separate univariate analysis. For each agent, the sales manager now has *several* variables; he now has *multivariate* data.

Simple Regression and Correlation

The first additional data to be considered are the scores of aptitude tests given each agent at the time of hiring. The test scores and sales figures for each agent are shown in Table 7-2. The two variables, test score and sales, are basically different in nature and in the roles they will play in this analysis. We wish to predict sales, and we plan to base that prediction on an aptitude test. In other words, we believe sales to be at least partially explained by aptitude. In the terminology of regression analysis, this leads us to identify sales as a *dependent variable* and aptitude test score as an *independent variable*. The graphing convention is to label the dependent variable Y and plot it on the vertical axis while labeling the independent X and plotting it on the horizontal axis. A plot of sales and aptitude test

TABLE 7-2   Aptitude test scores and life insurance sales for a sample of 10 sales agents

Agent	Aptitude test score (X)	Sales (Y), millions of dollars
1	67	.62
2	55	.81
3	58	.94
4	85	1.03
5	76	1.15
6	82	1.19
7	77	1.31
8	85	1.45
9	94	1.68
10	91	1.92

scores according to this convention is shown in Figure 7-2.

The plot of multivariate data is sometimes called a *scatter diagram*. Although the data are in fact scattered, an apparent relationship can be observed: low sales values tend to be associated with low test scores and high sales are associated with high scores. Notice that we avoid the temptation to suggest that high test scores *cause* high sales. Although some multivariate data may have underlying cause-and-effect relationships, many (including aptitude tests and insurance sales) do not. Success on tests and in sales performance may be due to other factors that are not easily measured or available for analysis. It makes no difference; if a second variable is closely related to the first and facilitates an improved estimation process, we do not care whether it is causative or not.

The scatter diagram also shows that $\overline{Y}$ may be a reasonable estimator of Y for aptitude test scores around 80, but it is consistently high for low test scores and consistently low for high test scores. Obviously, a line representing sales estimates that takes into consideration aptitude test scores should be drawn sloping upward from the lower left to the upper right. Such a line is shown in Figure 7-3.

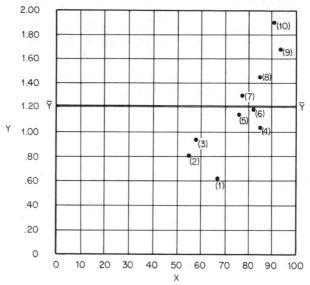

Figure 7-2 Scatter diagram of sales Y versus aptitude test score X.

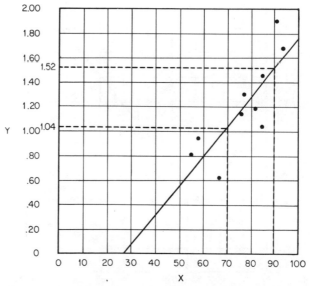

Figure 7-3 Regression line for sales Y versus aptitude test score X.

Now a point estimate of Y is dependent upon an associated value of X. For example, if an agent scored 70 on the aptitude test, we would estimate his or her sales to be 1.04 or \$1,040,000; for a test score of 90, the estimate would be 1.52 or \$1,520,000.

While it is apparent that the line in Figure 7-3 fits the data points better than the horizontal line in Figure 7-2, it is not clear how one constructs the line of *best* fit. In the form of two-variable analysis called *simple linear regression*, the line is placed so that the sum of the squares of the deviations is minimized. The line so formed is called the *regression* or *least-squares line*. Deviations from the regression line are usually measured parallel to the axis of the dependent variable, as shown in Figure 7-4, although it is interesting to note that squared deviations from the regression line are also minimized when measured parallel to the axis of the independent variable.

Like all straight lines, the regression line takes the form

$$Y = a + bX \tag{7-1}$$

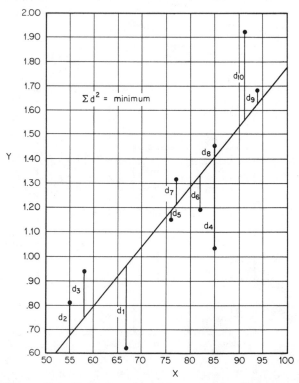

Figure 7-4 Deviations from the regression line.

where a is the Y intercept and b represents the slope or change in Y per unit increase in X. The values of a and b in the regression-line formula can be computed manually, but since they are included in the output of almost all computer regression programs, we shall not discuss the manual solution.

The Computer Analysis   The data from Table 7-2 were entered in a simple regression computer program. Aptitude test scores were identified as an independent variable and indexed as variable 1 (the program used, like most such programs, identifies variables by numbers instead of letters). Sales in millions of dollars were identified as the dependent variable and indexed as variable 2. The explanation of the resultant output is numerically keyed to Figure 7-5.

①   *Means and Standard Deviations.*   The program first calculates the mean and standard deviation for each variable. We recognize the mean and standard deviation of sales (index 2) from the univariate analysis; the mean and standard deviation of aptitude test scores (index 1) have no immediate use but are routinely furnished by the program.

②   *Parameters of the Regression Line.*   The values under the column heading B are the regression-line coefficients of the corresponding variable index.[2] Since there is no variable indexed 0, that coefficient, $-.646942$, is actually the constant a. And since index 1 refers to the X variable, that coefficient, $-.0241161$, is b, the coefficient of X. The regression-line equation can now be written as

$$\hat{Y} = -.646942 + .0241161X$$

The regression-line equation is interpreted as "sales in millions of dollars are equal to .0241161 times the aptitude test score minus .646942." In another sense, a represents the predicted sales for a test score of zero and b suggests that a sales increase of .0241161, or \$24,116.10, is associated with each one-point increase in test score.

Since a and b uniquely define the regression line, they are considered to be *parameters* just as the mean and standard deviation are parameters that uniquely define a normal distribution. However, since a and b were computed from sample data, they are technically *statistics* which serve as estimators of parameters. The true value of each parameter is somewhere in a normal distribution (t distribution for small samples) that has the

[2]The letter B actually represents the Greek letter β, which, properly subscripted, is often used to represent parameters in multiple regression. Thus, $Y = a + bX$ can be written $Y = \beta_0 + \beta_1 X$. Also, this computer program uses scientific notation for very small numbers; 2.41161 E-02 means $2.41161 \times 10^{-2} = 2.41161 \times .01 = .0241161$.

```
 ①  INDEX              MEANS               STANDARD DEVIATIONS
    1                  77                  13.2665
    3                  1.21                .395811


 ②  INDEX              B                   STD. ERROR      T-RATIO
    0                  -.646942            .484554         -1.33513
    1                  2.41161E-02         6.21049E-03      3.88313


 ③  R-SQUARED= .653359                     R= .808306


 ④  CORRELATION COEFFICIENTS
    1.                 .808308

    .808308            1.


 ⑤  ACTUAL             PREDICTED           RESIDUAL
    .62                .968839             -.348839
    .81                .679445             .130555
    .94                .751794             .188206
    1.03               1.40293             -.37293
    1.15               1.18588             -3.58841E-02
    1.19               1.33058             -.140581
    1.31               1.21                9.99997E-02
    1.45               1.40293             4.70705E-02
    1.68               1.61997             6.00257E-02
    1.92               1.54763             .372374


 ⑥  STAND. ERROR OF EST.= .247175                          D.F.= 8
```

Figure 7-5   Computer printout of a simple regression.

statistic (a or b) as a mean and a standard deviation as shown in the column headed STD. ERROR.

The value found in the T-RATIO column is simply the estimated parameter divided by its standard error. The t ratio facilitates a quick check of the hypothesis that the actual parameter value is zero. For example, at the .05 level of significance for eight degrees of freedom (10-2, for the two estimated parameters a and b), and a two-tail test, the critical t values are ± 2.306 (from Appendix G). Since the t ratio for a falls between these values, the hypothesis that $a = 0$ cannot be rejected; however, the t ratio for b is outside the critical t values and the hypothesis that $b = 0$ is rejected. These tests are illustrated in Figure 7-6.

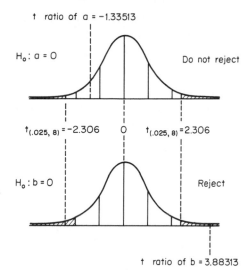

t ratio of a = −1.33513

H₀: a = 0   Do not reject

$t_{(.025,\ 8)} = -2.306$ 0 $t_{(.025,\ 8)} = 2.306$

Figure 7-6 Hypothesis tests   H₀: b = 0   Reject
of *a* and *b*.

t ratio of b = 3.88313

Whether *a* equals zero or not is of little consequence. The sales manager does not expect to hire any person with a test score near zero (the score associated with sales of *a*); besides, predictions of *Y* become unreliable when based on values of *X* appreciably outside the observed range of *X*. In this case, we would be reluctant to predict on the basis of a test score below 50.

The test of *b* holds greater importance. If we cannot say that *b* is significantly different from zero, we may as well abandon the regression model and use *Ȳ*, the mean of the univariate data, which, as a horizontal line, does have a slope of zero. Since there is only a 5 percent chance that *b* is zero, we continue this analysis.

③ *The Strength of the Regression Line.* The regression line explains the relationship between variables but says nothing about the *strength* of that relationship; that is, how well variation in the dependent variable is explained by variation in the independent variable. There are two closely related measures of this strength, the *coefficient of determination* and the *correlation coefficient*. Since the coefficient of determination is more easily defined in terms we have already discussed, we shall address it first.

The coefficient of determination is designated r^2 (R-SQUARED on the computer printout, which does not use lowercase letters) and is defined as the ratio of explained variance to total variance:

$$r^2 = \frac{\text{explained variance}}{\text{total variance}} \tag{7-2}$$

Explained variance can never be less than zero or more than total variance. This suggests a range of 0 to 1 for the coefficient of determination:

$$0 \leq r^2 \leq 1$$

Explained variance refers to the variation in Y that can be attributed to variation in X; in other words, explained variance is the variation in Y explained by the regression line. This interpretation can be understood more easily with a graphical analysis as shown in Figure 7-7. Point Y in Figure 7-7 represents a level of sales associated with aptitude test score X. Without X, we expect Y to be at $\overline{Y}$. The deviation $Y - \overline{Y}$ can be thought of as a total deviation. Total deviation leads to total variance:

$$\text{Total variance} = \frac{\Sigma(Y - \overline{Y})^2}{n - 1} \tag{7-3}$$

which we recognize as the variance of Y from which we determined s_Y in the univariate analysis.

When the values of X are considered, we expect Y to be at $\hat{Y}$ rather than $\overline{Y}$. The difference between $\hat{Y}$ and $\overline{Y}$ is explained by the regression line. Thus, the *explained* deviation $\hat{Y} - \overline{Y}$ leads to explained variance:

$$\text{Explained variance} = \frac{\Sigma(\hat{Y} - \overline{Y})^2}{n - 2} \tag{7-4}$$

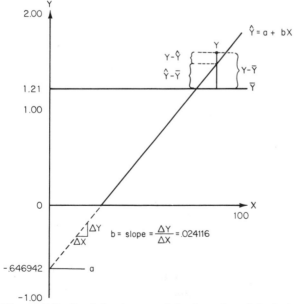

Figure 7-7 Explained, unexplained, and total deviations.

It is important to note that total variance is based on one estimated parameter $\overline{Y}$ and loses only one degree of freedom in the denominator while explained variance loses two for the two estimated parameters, a and b, required to compute $\hat{Y}$. The formulas for total and explained variance can now be substituted into formula (7-2).

$$r^2 = \frac{\Sigma(\hat{Y} - \overline{Y})^2/(n - 2)}{\Sigma(Y - \overline{Y})^2/(n - 1)} \qquad (7\text{-}5)$$

The third deviation shown in Figure 7-7, $Y - \hat{Y}$, will be discussed later.

For the insurance sales problem, the r^2 of .653359 means that slightly over 65 percent of the variation in sales is explained by variation in aptitude test score.

While the correlation coefficent does not measure explanatory power as simply or directly as the coefficient of determination, it is more commonly used to indicate the strength of the relationship between variables. One reason for the appeal of the correlation coefficient in simple regression is its ability to show whether the relationship between variables is direct or inverse. As the square root of the coefficient of determination, the correlation coefficient may be either positive or negative:

$$r = \pm \sqrt{\frac{\Sigma(\hat{Y} - \overline{Y})^2/(n - 2)}{\Sigma(Y - \overline{Y})^2/(n - 1)}}$$
$$= \pm \sqrt{r^2} \qquad (7\text{-}6)$$

For the minimum r^2 value of zero, r is also zero; but for the maximum r^2 of 1, r may be either $+1$ or 1. The range of r is therefore

$$-1 \leqslant r \leqslant 1$$

By convention, the negative root is used to identify an *inverse* relationship (one in which the dependent and independent variables move in opposite directions), and the positive root is used to identify *direct* relationships (those in which the variables move together). Since insurance sales increase as aptitude test scores increase, the relationship is direct and the correlation coefficient is $+ \sqrt{.653359} = +.808306$. Examples of other relationships are shown in Figure 7-8.

The relationship between variables can also be inferred from the sign of b in the equation of the regression line. In simple linear regression, r and b always have the same sign.

④   *The Correlation Coefficient Matrix.*   In addition to showing R-SQUARED and R, the computer printout also shows a matrix labeled

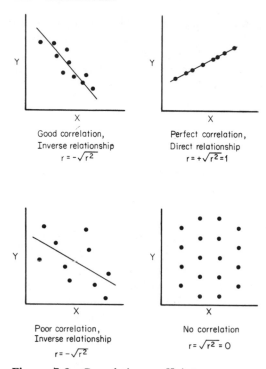

Good correlation,
Inverse relationship
$r = -\sqrt{r^2}$

Perfect correlation,
Direct relationship
$r = +\sqrt{r^2} = 1$

Poor correlation,
Inverse relationship
$r = -\sqrt{r^2}$

No correlation
$r = \sqrt{r^2} = 0$

Figure 7-8 Correlation coefficients.

Correlation Coefficients.[3] Such a matrix has greater significance in the multiple-regression analysis undertaken later, but it does illustrate two interesting points that also apply to simple regression. The first point is that a data set correlates perfectly with itself. That is, if X (index 1) values were plotted on both axes, the correlation coefficient would be 1. This is reflected by the coefficient 1 in the upper left-hand corner of the matrix (the upper left-hand corner is the intersection of the row and column for index 1). Similarly, Y (index 2) values correlate perfectly with themselves as indicated by the 1 at the intersection of row 2 and column 2.

The second point is that the correlation coefficient is the same regardless of the designation of the dependent variable. (This was hinted at earlier in the statement that the least-squares line relects minimum squared deviations from either axis.) Thus, the correlation of Y (variable 2) with X (variable 1), found at the intersection of row 2 and column 1, is the same as the correlation of X with Y, found at the intersection of row 1 and column 2; both have coefficients of .808308.

[3]The very slight difference between the value of r in these two listings is attributed to rounding in arithmetic operations internal to the computer.

(5) *Actual, Predicted, and Residual Values.* The table of actual, predicted, and residual values shows, in order, Y, $\hat{Y}$, and $Y - \hat{Y}$. The actual values are the observed values of Y entered into the program from Table 7-2. The predicted values are the estimates $\hat{Y}$ based on the regression line and the corresponding value of X. For example the predicted sales for agent 4 are

$$\hat{Y} = a + bX$$
$$X = 85 \quad \hat{Y} = -.646942 + .0241161(85)$$
$$= -.646942 + 2.049869$$
$$= 1.40293$$

The residual value, $Y - \hat{Y}$, is the third deviation illustrated in Figure 7-7. The residual represents that part of the total deviation *not* explained by the regression line. It leads to the computation of what is called *unexplained variance*:

$$\text{Unexplained variance} = \frac{\Sigma(Y - \hat{Y})^2}{n - 2} \tag{7-7}$$

In the manual solution of a linear-regression problem, it is desirable to reduce hand computation as much as possible. Since unexplained variance has several uses, and since explained variance is used only in the computation of the coefficient of determination, an algebraic substitution is often made to use unexplained variance in the coefficient of determination and to avoid using explained variance completely. The substitution is based on the fact that unexplained plus explained variance gives total variance:

Unexplained variance + explained variance = total variance   (Given)

Explained variance = total variance − unexplained variance

                                                        (Subtraction)

$$r^2 = \frac{\text{explained variance}}{\text{total variance}} \qquad\qquad \text{[Equation (7-5)]}$$

$$= \frac{\text{total variance} - \text{unexplained variance}}{\text{total variance}} \qquad\qquad \text{(Substitution)}$$

$$= \frac{\text{total variance}}{\text{total variance}} - \frac{\text{unexplained variance}}{\text{total variance}} \qquad\qquad \text{(Division)}$$

$$= 1 - \frac{\text{unexplained variance}}{\text{total variance}} = 1 - \frac{\Sigma(Y - \hat{Y})^2/(n - 2)}{\Sigma(Y - \bar{Y})^2/(n - 1)} \tag{7-8}$$

and
$$r = \pm\sqrt{1 - \frac{\Sigma(Y - \hat{Y})^2/(n - 2)}{\Sigma(Y - \bar{Y})^2/(n - 1)}} \tag{7-9}$$

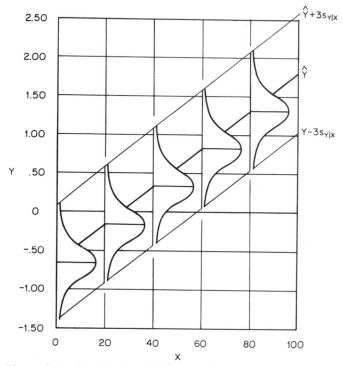

Figure 7-9 Distribution of Y about the regression line.

(6) *The Standard Error of the Estimate.* The residuals and unexplained variance have a much more important role in regression analysis than merely providing an alternate formula for r and r^2. Unexplained variance is actually the variance of Y with respect to $\hat{Y}$, just as total variance is the variance of Y with respect to $\overline{Y}$. And just as we illustrated the distribution of Y about $\overline{Y}$ with a normal curve in Figure 7-1, we can now illustrate, in Figure 7-9, the distribution of Y about $\hat{Y}$ as a sort of three-dimensional normal curve moving along the regression line.

Variance is helpful in understanding the coefficients of determination and correlation, but the square root of variance, the standard deviation or standard error, is more useful in the estimation process. When we estimated on the basis of $\overline{Y}$, we used s_Y, the standard deviation of Y. Now, we have explained the variance between $\overline{Y}$ and $\hat{Y}$, and we shall estimate on the basis of $\hat{Y}$. For an interval estimate, however, we need a measure of variation. In this case, the appropriate measure is $s_{Y|X}$, *the standard error of the estimate*, which is the square root of unexplained variance:[4]

---

[4]The subscript in $s_{Y|X}$ is read as "Y given X." Thus, $s_{Y|X}$ is the standard error of Y given X. This subscript is sometimes written Y.X or even YX. We prefer Y|X for its consistency with probability notation.

$$s_{Y|X} = \sqrt{\frac{\Sigma(Y - \hat{Y})^2}{n - 2}} \qquad (7\text{-}10)$$

On the computer printout in Figure 7-5, the standard error of the estimate is .247175. On the same line as the standard error, the printout indicates the degrees of freedom (D.F.) used in the computation. In this case, ten observations minus two estimated parameters gives eight degrees of freedom.

Estimates of Y Given X   Earlier, we estimated sales from aptitude test scores and a graphical representation of the regression line. Now, with a mathematical expression for the regression line, we can compute a point estimate of Y given X in exactly the same manner as predicted values of Y were computed. For example, the sales of an agent who scores 70 on the aptitude test are estimated as

$$\hat{Y} = a + bX$$
$$= -.646942 + .0241161(70)$$
$$= 1.041185$$

or approximately $1,040,000.

The determination of a standard error of the estimate now enables us to build a confidence interval about that point estimate. For example, a 90 percent confidence interval estimate of sales for the same agent is given by

$$\hat{Y} \pm t_{(\alpha/2, \mathrm{df})}\, s_{Y|X} = \hat{Y} \pm t_{(.05, 8)} s_{Y|X}$$
$$= 1.041185 \pm 1.860(.247175)$$
$$= .581439 \text{ to } 1.500931$$

or approximately $580,000 to $1,500,000.

The range of this estimate, approximately $920,000, can be seen in Figure 7-10 to be an improvement over the estimate based on univariate data, but it is still a very imprecise value. Let us see if further improvement is possible.

Multiple Regression and Correlation

The extension of simple regression techniques to two or more independent variables is called *multiple regression*. In an effort to refine the predictive power of the regression model, the sales manager decides to add a second independent variable and perform a multiple regression. The new variable is one suggested earlier, the income of prospective clients. The

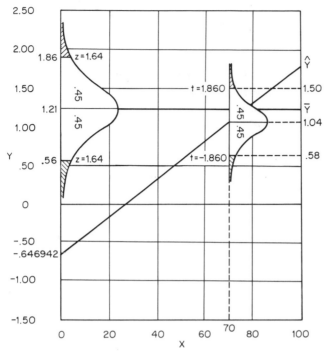

Figure 7-10 Comparison of univariate and multivariate interval estimates of Y.

measure of this variable is the median family income, in thousands of dollars, for each agent's sales area. Median incomes, along with the original variables, are shown in Table 7-3.

Since median income is an independent variable, we shall identify it as an X variable. The three variables and their respective computer program indexes are now

X_1, aptitude test score   (index 1)
X_2, median family income   (index 2)
Y , insurance sales   (index 3)

In univariate analysis, the one-dimensional line used an estimator of *no* dimensions, a point. In simple linear regression, the two-dimensional graph used an estimator of *one* dimension, a line. Now, in multiple regression, we shall depict the three variables in three dimensions. How shall we represent the estimator? Since the estimator seems to have one less dimension than the model, we would expect a *two*-dimensional figure, a plane. The three-dimensional model does in fact possess a *regression plane* by which we shall estimate sales Y based on aptitude test score X_1 and median family income X_2.

The general formula for the regression plane is

$$\hat{Y} = a + b_1 X_1 + b_2 X_2 \qquad (7\text{-}11)$$

where $a = Y$ intercept

$b_1 =$ slope of the line formed by the intersection of the regression plane and the plane formed by X_1 and Y

$b_2 =$ slope of the line formed by the intersection of the regression plane and the plane formed by X_2 and Y

The regression plane and the estimated sales for an agent with an aptitude test score of 70 operating in an area where the median family income is $15,000 are shown in Figure 7-11.

The Computer Analysis The computer analysis of a multiple-regression problem follows the same format as that for simple linear regression. Let us examine the multiple-regression printout of the insurance sales problem shown in Figure 7-12, concentrating on areas where there is an interpretation unique to multiple regression.

(1) *Means and Standard Deviations.* The means and standard deviations of sales (index 3) and aptitude test score (index 1) are, of course, unchanged. The new variable, median family income (index 2), shows a mean of 15.4 ($15,400) and a standard deviation of 1.658 ($1658).

(2) *Parameters of the Regression Plane.* Again, index 0 refers to a, the Y intercept. Now, however, the Y axis represents not only zero aptitude but zero median income as well! Obviously, this is a theoretical point only, and its negative value of -2.48024 has no particular significance.

TABLE 7-3 *Aptitude test scores, median family incomes, and life insurance sales for a sample of 10 sales agents*

Agent	Aptitude test score (X_1)	Median family income (X_2), thousands of dollars	Sales (Y), millions of dollars
1	67	12.5	.62
2	55	16.6	.81
3	58	16.1	.94
4	85	14.6	1.03
5	76	15.3	1.15
6	82	13.4	1.19
7	77	16.5	1.31
8	85	15.1	1.45
9	94	15.6	1.68
10	91	18.3	1.92

Index 1 refers to b_1, the coefficent of X_1 in the formula for the regression plane. Even though b_1 is a very small value (.0229656), it has a very large t ratio, an indication that a null hypothesis of $b_1 = 0$ should be rejected. Since b_1 is positive, we expect the regression plane to slope upward with respect to the plane of X_1Y, an expectation confirmed in Figure 7-11.

Index 2 identifies b_2, the coefficient of X_2. The value of b_2, .124798, suggests a positive slope in the X_2Y plane. The t ratio of b_2 is also large, confirming that the nonzero slope of b_2 is not just due to the chance involved in drawing this particular sample of agents.

It can also be seen that b_2 is larger than b_1 and that the slope in the X_2Y plane is steeper than it is in the X_1Y plane. While this observation is logical in general terms, it should be noted that the scales of Y, X_1, and X_2 are different on the graph. Difference in scale will cause distortion in the slope and make visual comparisons difficult. For example, if Y and X_1 were both on a scale of 0 to 100, the slope of .0229656 would be virtually indistinguishable from a zero-sloped horizontal line.

The values of the regression-plane parameters can now be substituted into equation (7-11):

$$\hat{Y} = -2.48024 + .0229656X_1 + .124798X_2$$

③   *Strength of the Regression Plane.*   When the relationships between all independent variables and the dependent variable are considered simul-

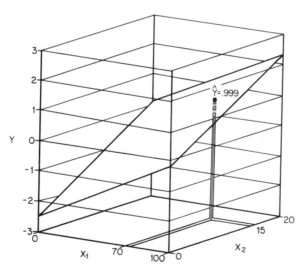

Figure 7-11 Regression plane for sales Y vs. aptitude test score X_1 and median family income X_2.

① INDEX   MEANS   STANDARD DEVIATIONS
   1   77   13.2665
   2   15.4   1.658
   3   1.21   .395811

② INDEX   B   STD. ERROR   T-RATIO
   0   -2.48024   .436062   -5.68781
   1   2.29656E-02   3.09347E-03   7.42389
   2   .124798   2.47522E-02   5.04191

③ R-SQUARED= .925156   R= .96185

④ CORRELATION COEFFICIENTS
   1.   .073775   .808308

   .073775   1.00001   .57956

   .808308   .57956   1.

⑤ ACTUAL   PREDICTED   RESIDUAL
   .62   .61843   1.56951E-03
   .81   .854516   -4.45156E-02
   .94   .861013   7.89866E-02
   1.03   1.29389   -.263886
   1.15   1.17456   -.024555
   1.19   1.07523   .114768
   1.31   1.34728   -3.72782E-02
   1.45   1.35629   9.37147E-02
   1.68   1.62538   .054625
   1.92   1.89343   2.65679E-02

⑥ STAND. ERROR OF EST.= .122783   D.F.= 7

Figure 7-12   Computer printout of a multiple regression.

taneously, the coefficient of determination is called the *coefficient of multiple determination* and is labeled R-SQUARED. Its interpretation is the same as that of r^2: the percentage of variance in the dependent variable explained by the regression. In this case, we see that over 92.5 percent of the variance in Y is explained by the regression or, in other words, over 92.5 percent of the variance in sales is explained by variance in aptitude test scores and median family incomes. This is an appreciable improvement over the 65 percent explained by aptitude test scores alone in simple regression.

The square root of R^2, the coefficient of multiple determination, is R, the *multiple correlation coefficient*. Unlike the simple correlation coefficient, R is always a positive square root. The reason for ignoring the negative root is that multiple relationships are not strictly direct or inverse. It is possible for one independent variable to have an inverse relationship with the dependent variable while a second one has a direct relationship. In such cases it is not meaningful to describe the multiple relationship as either direct or inverse; consequently, sign is ignored for the multiple correlation coefficient

$$R = \sqrt{R^2} = \sqrt{.925156} = .96185$$

④ *The Correlation Coefficient Matrix.* The matrix of correlation coefficients is particularly interesting in multiple regression. Since the computer-generated matrix is unlabeled, we have reproduced the matrix, with column and row headings, as Table 7-4.

Although this is a problem in *multiple* regression, Table 7-4 consists of *simple* correlation coefficients. The diagonal of 1s will be recognized as the correlation coefficients of X_1 with X_1, X_2 with X_2, and Y with Y. The upper right and lower left corner values of .808308 are also familiar as the simple correlation coefficients of X_1 with Y and Y with X_1, as determined earlier in simple linear regression. By similar reasoning, we see that the simple correlation coefficient of X_2 with Y and Y with X_2 is .57956. Squaring this value gives a simple coefficient of determination of .33589, meaning that approximately 33.6 percent of the variation in Y (sales) is explained by X_2 (median family income). It is interesting that two variables with relatively modest simple correlations should combine to give very strong multiple correlation.

The remaining coefficients in the matrix are also simple correlation coefficents of the row variable with the column variable but have a completely different meaning to an analyst. If X_1 and X_2 correlate well with each other, it may be that one or the other is unnecessary. For example, if we were to predict the annual income of lawyers from age and years of experience, we would expect a high simple correlation coefficient between age and experience, since young lawyers are likely to have less experience

TABLE 7-4 Correlation coefficients

Variable	Index	X_1 1	X_2 2	Y 3
X_1	1	1	.073775	.808308
X_2	2	.073775	1	.57956
Y	3	.808308	.57956	1

than older ones. Strong correlation between independent variables gives rise to a situation called *multicollinearity*. Multicollinearity is undesirable because the redundant variable usually adds little to the multiple correlation coefficient and may actually increase the standard error of the estimate by reducing the degrees of freedom in the denominator. In this case, X_1 and X_2, with a simple correlation coefficient of .073775, are almost totally unrelated (a coefficient of zero represents no correlation at all), and both should be retained.

⑤ *Actual, Predicted, and Residual Values.*   In multiple regression, predicted values of the dependent variable are also computed by substituting observed values of the independent variables into the regression equation. That equation is now the equation of a plane instead of a line, of course. For comparison with the linear-regression model, we again compute predicted sales for observation (agent) 4:

$$\hat{Y} = a + b_1X_1 + b_2X_2$$
$$X_1 = 85 \qquad X_2 = 14.6$$
$$\hat{Y} = -2.48024 + .0229656(85) + .124798(14.6)$$
$$= -2.48024 + 1.95208 + 1.82205$$
$$= 1.29389$$

⑥ *The Standard Error of the Estimate.*   Even with the introduction of a second independent variable, there is still some unexplained variance in sales. As in simple linear regression, the square root of unexplained variance is called the *standard error of the estimate*, although the notation is slightly different to reflect the dependence upon more than one independent variable:[6]

$$s_{Y|X_1,X_2} = \sqrt{\frac{\Sigma(Y - \hat{Y})^2}{n - 3}} \qquad (7\text{-}12)$$

The standard error of the estimate for this multiple regression is .122783, less than one-half the standard error of the estimate for the simple regression. This reduction was achieved in spite of the fact that fewer degrees of freedom in the denominator tend to increase the standard error. In this case, the seven degrees of freedom are based on ten observations and the three estimated parameters, a, b_1, and b_2, required to compute each $\hat{Y}$.

---

[6]We have extended the simple-linear-regression notation to multiple regression. It is perhaps more common to see the standard error of the estimate in multiple regression identified by index numbers than by the actual variables. In such notation, $s_{Y|X_1,X_2}$ becomes $s_{3|12}$ or $s_{3.12}$.

Estimates of Y Given X_1 and X_2   The earlier graphical estimation of sales for an aptitude test score of 70 and median family income of $15,000 was largely illustrative. It would be very unusual to attempt an estimate from a three-dimensional graph and impossible to make one for four or more dimensions. Again, point estimates are really computed like predicted values. For $X_1 = 70$, $X_2 = 15$, the point estimate of Y is

$$\hat{Y} = a + b_1 X_1 + b_2 X_2$$
$$= -2.48024 + .229656(70) + .124798(15)$$
$$= .999322$$

or approximately $1,000,000.

A 90 percent confidence interval is found similarly to the simple regression interval:

$$\hat{Y} \pm t_{(\alpha/2,\,\mathrm{df})}\, s_{Y|X_1,X_2} = \hat{Y} \pm t_{(.05,\,7)} s_{Y|X_1,X_2}$$
$$= .999322 \pm 1.895(.122783)$$
$$= .766648 \text{ to } 1.231996$$

or approximately $770,000 to $1,230,000.

The interval estimate of Y given X_1 and X_2 is difficult to depict graphically. In simple regression, we added a third dimension to show the distribution of Y about the regression line. In multiple regression, Y is also

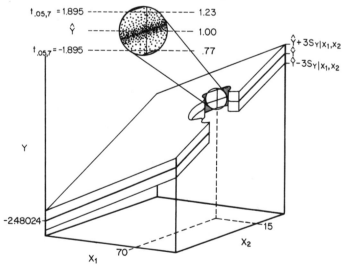

Figure 7-13   Interval estimate of sales Y based on aptitude test score X_1 and median family income X_2.

TABLE 7-5 *Estimates of insurance sales (Y) to the nearest thousand dollars*

Independent variable(s)	Point	.90 interval	Interval range
None	1,210,000	561,000–1,859,000	1,298,000
Aptitude test score (70)	1,041,000	581,000–1,501,000	920,000
Aptitude test score (70) and median family income ($15,000)	999,000	767,000–1,232,000	465,000

distributed normally (*t*, for small samples) about the regression plane, but we cannot add a fourth dimension. Instead, we try to visualize the regression plane as being approximately six standard errors of the estimate thick with a *density* that is normally distributed, very dense at the center and less dense toward the surfaces. The 90 percent confidence interval estimate of *Y* given X_1 and X_2 is shown this way in Figure 7-13.

It is now interesting to observe the improvement in the estimation of sales as independent variables have been added. A comparison of the point and interval estimates for univariate, simple linear regression, and multiple regression is shown in Table 7-5.

OTHER TOPICS IN REGRESSION ANALYSIS

Although the coverage of regression and correlation in this chapter is sufficient for normal business applications, some managers may have access to computer programs that print the results of more sophisticated regression techniques. Complete explanations of these techniques are beyond the scope of this book, but a few brief descriptions will enable the curious reader to determine the need for further study in a more specialized text.

Partial Coefficients

There are several coefficients in multiple regression that are identified as *partial*. For example, the coefficients b_1 and b_2 in the formula for the regression plane are sometimes called *partial regression coefficients*. Of greater interest, however, is the *coefficient of partial determination* and its square root, the *partial correlation coefficient*.

Like other coefficients of determination, the coefficient of partial determination measures the percentage of variance explained by a variable. But where the coefficient of determination measures the variance explained by all independent variables, the coefficient of partial determination measures only the added explanation that results from the introduction of a new variable. For example, if X_1, X_2, and X_3 are independent variables, the coefficient of partial determination of X_3 gives the percentage of variance unexplained by X_1 and X_2 that is explained by X_3. Coefficients of partial determination are useful when choosing which of several possible variables to include in an analysis.

Stepwise Regression

The concept of partial coefficients leads directly to a procedure for selecting independent variables called *stepwise regression*. In stepwise regression, independent variables are introduced one at a time, in the order of their coefficient of partial determination, until the coefficient of multiple determination is maximized. It is possible for variables to be dropped during such an analysis also, but the result is always a maximum coefficient of multiple determination. Stepwise regression is usually a separate program from multiple or simple linear regression in statistical program libraries.

Autocorrelation

Earlier, we mentioned the possibility of multicollinearity, in which independent variables are correlated with each other, giving, perhaps, a false impression that both were correlated to the dependent variable. A similar problem, but harder to detect, is that of *autocorrelation*, in which the dependent variable is correlated to itself. Autocorrelation implies that the value of the dependent variable is determined, at least partially, by its own previous value(s). Sometimes a cyclical trend will show up on a scatter diagram, but this diagnosis will work only for simple linear regression. In multiple regression, we rely on a measure called the *Durbin-Watson statistic*. When the Durbin-Watson statistic is given on a computer printout, it must be interpreted with the aid of a special table usually found only in advanced statistics textbooks.

A FINAL CAUTION

Multiple regression is one of the most powerful tools in the quantitatively oriented manager's inventory. It is based on historical data that are often readily available, and the analyst need not justify the inclusion of independent variables rationally; if they add to the explanatory power of the model, they are welcome.

In these days of mandatory data collection and reporting laws and with the aid of modern computers, multiple regression seems to be a natural analytical technique. A few rules must be observed, however: (1) The relationships must be linear. A relationship based on the square or reciprocal of an independent variable is not linear and must not be included. (2) The regression is not valid outside the limits of observed data. There is a classic example of this rule that makes its point rather emphatically: if 1 milligram of a heart stimulant increases pulse by five beats per minute, what pulse is predicted by 100 milligrams? Assuming a pulse of 70 with no stimulant, we compute

$$\hat{Y} = a + bX$$
$$= 70 + 5(100)$$
$$= 570 \text{ heartbeats per minute!}$$

Obviously there is a limited range to X (milligrams of stimulant) and we can only hope that heartbeat levels off at some point instead of dropping to zero.

The restriction on the range of independent variables actually works to our advantage in some cases. For example, consider the scatter diagram in Figure 7-14. Certainly, this is not a linear relationship. However, by restricting the independent variable to values between X_a and X_b, we may be able to approximate the relationship with a linear model.

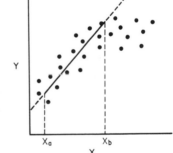

Figure 7-14 Limited linear approximation of a nonlinear, relationship .

Finally, predictive ability should not be confused with manipulative ability. The temptation to influence the dependent variable by altering an independent variable is strong. Particularly in economic models, where the number of variables is large and the relationships complex, it is not reasonable to assume that, say, disposable income can be raised by increasing the wages of automobile workers, even though the two may have a positive simple correlation coefficient. There are interactions with other variables, such as the price of automobiles, that may cause a reaction quite opposite to that desired.

PART 3
Quantitative Methods

*Quantitative management techniques have become
increasingly popular in the last 20 years. Most business schools now require their students to develop the
basic quantitative skills necessary to communicate and
work with industrial engineers, operations researchers, and other technical specialists. When used correctly, the ability to develop and analyze a quantitative
model often improves the decision-making process.*

*Another reason for the increased use of quantitative
methods is the rapid development in computer
technology. Modern electronic computers can handle
cumbersome and complex calculations quickly and
efficiently. Most computer manufacturers have
software packages available with the solution
procedures for various quantitative techniques already
programmed and ready to use. Furthermore, the
technology has developed to the point where even
relatively small businesses can afford this type of
computer capability.*

*Since the computational work is often done by
computer, a manager need only have a basic
understanding of what the quantitative methods are
all about, what they do, and when they can be
successfully used. This is the approach that is taken in*

describing the quantitative management techniques presented in Chapters 8 to 12. Emphasis is on concepts, assumptions, and potential applications rather than computational details. Unlike the statistics chapters in Part 2, which build on each other, each chapter in Part 3 is an independent discussion of an important quantitative management technique.

CHAPTER 8

Linear Programming

Linear programming (LP) is a method of analysis that has found many useful applications in business. For example, many problems in the area of production scheduling, financial planning, purchasing, and marketing strategy can be solved by linear programming. In general, problems that lend themselves to this type of analysis involve finding the best way to schedule the use of some set of resources that is limited in supply. Depending on the nature of the problem, the resources might be money, man-hours, machine capacities, warehouse space, or some other factor whose availability affects the decision at hand.

The word *programming* in this sense does not refer to a computer, as one usually associates the term, but implies that a specific set of mathematical procedures is used to solve the problem. The adjective *linear* refers to a relationship that must exist between the elements of the problem, which will be discussed later.

A linear-programming analysis consists essentially of expressing the problem situation as a series of simple mathematical equations or inequalities and finding the solution that best achieves some stated objective. In production scheduling, the problem might consist of finding the combination of products to process in some planning period so that total contribution to profit and overhead is maximized while at the same time the availability of man- and/or machine-hours is not exceeded. As another example, a marketing-strategy problem might consist of allocating a given budget to various advertising media so that maximum exposure to the product or service is achieved.

Even though the techniques of linear programming have been known for some time, only with the development of high-speed electronic computers has this method of analysis seen widespread applications in business. Most computer manufacturers now have software packages available that contain canned programs in which the procedures for solving linear-programming problems are already programmed into the computer. Since a computer can do the computational work, a manager need not be a mathematician to use linear programming successfully. What a manager must be able to do, however, is (1) recognize when linear programming is an appropriate method of analysis, (2) formulate the problem so that it can be solved (either by computer or manually), and (3) interpret the information that results and use it correctly.

WHEN IS LINEAR PROGRAMMING APPROPRIATE?

In many ways, knowing when a particular type of technique or method of analysis will be appropriate is an art. An important part of a decision maker's knowledge about a technique such as linear programming should be a feeling for when it will work. This essentially consists of understanding the assumptions inherent in the analysis and being able to relate them to the particular decision situation. It is an art because very rarely will the assumptions of the technique be completely consistent with the realities of the situation. Things are almost never black or white; a lot of gray areas exist in the real world. Good judgment is often required to determine whether the assumptions match the situation well enough for a meaningful analysis to result.

Nevertheless, there are some specific characteristics that a problem must have before linear programming should be considered as a method of analysis. Before looking at an example problem and the solution methodology, we identify and discuss these important characteristics.

Some Specific Things to Look For

First, the situation must be such that a single objective can be identified for the decision. For example, a purchasing manager might want to schedule the purchases of raw materials so that total cost is minimized. In the example cited earlier, an advertising manager might want to select advertising media so that maximum exposure to the product is attained. Alternatively, a production scheduler might want to schedule work on a certain piece of equipment or process in such a manner that total idle time is minimized.

In all these examples there are probably other factors that are also relevant, but one objective can be singled out as the most important or the most critical. Before implementing a solution based on an analysis that involves only the single objective, the decision maker will, of course, want to consider the other relevant factors as well. For example, the purchasing manager may want to consider the effect that the "optimal" purchasing plan would have on the company's working relationship with the vendors. This much is clear: decision situations that involve conflicting objectives of seemingly equal importance do not lend themselves well to linear-programming analysis.

Second, the decision situation must contain certain restrictions on the alternative solutions. These restrictions or constraints are typically the result of such things as budget limitations, sales commitments, material requirements, capacity limitations, and so on. For example, the development of the most economical purchasing plan might have to be done within certain contractual agreements that have already been made with the vendors. The advertising strategy would generally have to be developed within budget limitations. Production scheduling might have to be accomplished within the requirement that certain outputs be achieved each planning period. In many problems, there are a large number of constraints that must be dealt with.

The use of linear programming also implies that the objective and the constraints can be expressed as linear, or straight-line, equations. For example, if product A requires 5 minutes of processing time and sells for $1, a production and sales volume of 100 units of product A would require 500 minutes of processing time and generate $100 in sales. Similarly, if the production and sales volume were 1000 units, 5000 minutes of processing time would be needed and sales would be $1000. Thus, to say that there is a linear relationship means that unit data (cost, processing time, price, etc.) remain the same, regardless of the level of volume.

Sometimes the straight-line relationship is not a good description of reality. Often economies of scale result from large operations where, after a certain point, additional volume results in lower unit costs. This often happens in purchasing, where discounts in prices are offered when large quantities are ordered. In production, the situation could conceivably be reversed. At a certain level of operations, a set of facilities could be so taxed by an increase in volume that unit costs go up. In either case, unit data would be affected by volume, and therefore a linear relationship would not exist.

However, in certain cases linear programming might still be useful when the constraints are not strictly straight-line equations or inequalities. Often a straight line is a good approximation for a curve, especially over a limited segment of the curve. Thus, even when a constraint (or objective) faced by the decision maker is more accurately represented by a curved line, the portion of the curve in which the company operates might be approximated rather nicely by a straight line. In those cases, linear programming will probably provide a meaningful analysis.

Finally, it is helpful simply to be familiar with some general types of problems that have been solved by linear programming in the past. Certain classes of problems (some of which we have already mentioned) are known to be solvable by linear programming. A general knowledge of these may provide the first clue that linear programming is appropriate for the particular problem under study. Below are some additional types of problems that can be solved by linear programming.

1. *Blending problems.* Two or more ingredients must be blended into a product. The objective is to determine the quantity of each ingredient to be used so that the product exhibits the desired characteristics and total cost is minimized.

2. *Make or buy problems.* An item or group of items can either be purchased or produced. Given the unit purchase cost, unit production cost, quantities needed, and available production facilities, the objective is to determine which items (and how many) should be purchased and which produced so that total acquisition cost is minimized.

3. *Portfolio-selection problems.* Investment alternatives are selected so that the expected return is maximized and the portfolio is within acceptable limits of risk, diversity, available funds, and so on.

4. *Transportation problems.* A set of supply points (plants, warehouses) serves a set of destination points (retail outlets), each distribution channel having a unique cost. The objective is to distribute total supply to the destination points in such a manner that demand is met and total costs are minimized.

5. *Assignment problems.* A group of workers must be assigned to an equal number of tasks. Here the assignments are one for one. For example, six workers must be assigned to six different tasks. The objective is to make the assignments so that total cost is minimized or total benefit is maximized.

FORMULATING LINEAR-PROGRAMMING PROBLEMS

To illustrate the procedure for formulating and solving linear-programming problems, take a simple example. Suppose a furniture manufacturer produces a line of lamps in which two variations of a basic lamp can be made, a desk model and a floor model. The desk model generates $10

TABLE 8-1   Processing times for desk and floor lamps

Operation	Unit processing time, minutes		Man-hours available
	Desk model	Floor model	
Wiring	15	20	80 (4800 minutes)
Assembly	10	20	60 (3600 minutes)
Packaging	5	8	40 (2400 minutes)

per unit in contribution to profit and overhead, while the more expensive floor model contributes $15.

The two models are made on the same production line, and the production schedule for the coming week must be prepared. The problem is to develop the production schedule that makes best use of the resources available. Since the same processes are used to make both models, every desk model that is scheduled for production will use up some resources that would otherwise be used to make a floor model, and vice versa.

The production area consists of three separate work centers, where the wiring, assembly, and packaging are done. Table 8-1 shows the estimated time required to wire, assemble and package each type of lamp, as well as the man-hours available over the next week in each of the work centers. Additionally, we shall assume that no more than 100 floor models will be needed to satisfy demand and replenish inventories over the coming week.

Before going any farther, we should verify that we do, in fact, have a linear-programming problem here. The objective is to schedule for production that combination of desk and floor models that will maximize total contribution to profit and overhead. The availability of man-hours in each work center forms a definite constraint on the possible production quantities of each model. Also, the limitation on the demand for floor models affects the possible production schedule. Finally, the data in Table 8-1 indicate that the assumption of linearity is valid. For example, it will require 15 minutes to wire each desk lamp, regardless of whether the production run is for 1 or 100 lamps. This type of problem is often referred to as a *product-mix* problem, and it is one of the classic examples of the application of linear programming.

In order to formulate the problem in linear-programming format, the objective and constraints must be stated as mathematical equations or inequalities. The mathematical formulation forces the decision maker to be very explicit about what is to be achieved and what factors affect the decision. The objective is formulated as

$$\text{Maximize contribution} = \$10D + \$15F,$$

where D is the number of desk models to schedule for production and F is the number of floor models. These are called the *decision variables.*

The equation for the objective is referred to as the *objective function,* and in this case it implies that increases in production of the two models will result in higher total contribution. While this is true, there is a limit to the availability of man-hours in the three centers where the actual work is done. Therefore, the objective function is to be maximized subject to certain constraints. For the work center where the wiring is done, the constraint is formulated as

$$\left[\begin{array}{c} \text{Number of minutes needed to} \\ \text{wire combination of lamps} \\ \text{scheduled for production} \end{array} \right] \leq \left[\begin{array}{c} \text{number of minutes available} \\ \text{in the Wiring Department} \end{array} \right]$$

or

$$15D + 20F \leq 4800$$

For any production schedule (values for D and F) the left side of the inequality gives the total number of minutes required for wiring. For example, if the schedule called for the production of 150 desk models and 100 floor models, the total time needed for wiring would be $15(150) + 20(100) = 4250$ minutes. The constraint limits the production schedule to any combination of desk and floor models that can be wired within the 4800 minutes available in that work center for the coming week. The constraints imposed by the availability of assembly and packaging time are formulated in a similar manner:

$$10D + 20F \leq 3600 \quad \text{(Assembly)}$$
$$5D + 8F \leq 2400 \quad \text{(Packaging)}$$

Finally, the limited demand for floor models is expressed mathematically as

$$F \leq 100$$

Putting all this together, we can express the entire scheduling problem as the linear program

$$\text{Maximize contribution} = \$10D + \$15F$$

subject to

$$15D + 20F \leq 4800 \quad \text{(Wiring)}$$
$$10D + 20F \leq 3600 \quad \text{(Assembly)}$$
$$5D + \ 8F \leq 2400 \quad \text{(Packaging)}$$
$$F \leq \ 100 \quad \text{(Demand for floor models)}$$

The mathematical formulation is really no more than an exercise in logic. It results in an explicit, unambiguous statement of the problem. In fact, the formulation exercise alone probably provides some valuable insights into the problem. It happens, though, that the optimal solution to this particular problem can also be found rather easily. Before looking at the solution methods, it will be worth examining another formulation example.

Another Example

A company operates two separate plants in which it makes a product that it distributes to regional warehouses in three market areas. Both plants can ship to any of the three warehouses, but each distribution channel has a unique cost. The problem is to develop a distribution plan so that weekly production in the two plants is allocated to meet weekly demand in the three market areas and the total distribution cost is minimized. This is the basic format of the transportation problem mentioned earlier.

Table 8-2 gives the weekly plant capacities, weekly market demand, and unit cost to distribute from each plant to each regional warehouse. Notice that total weekly capacity in the two plants is exactly equal to total weekly demand in the three market areas. This certainly does not have to be the case to use linear programming, but we assume that it is so in order to keep the example as simple as possible.

In order to formulate the objective function and the constraints, the following notation will be used to identify the decision variables in the problem:

X_{P1-W1} = number of units to be shipped from Plant 1 to Warehouse 1
X_{P1-W2} = number of units to be shipped from Plant 1 to Warehouse 2
X_{P1-W3} = number of units to be shipped from Plant 1 to Warehouse 3
X_{P2-W1} = number of units to be shipped from Plant 2 to Warehouse 1
X_{P2-W2} = number of units to be shipped from Plant 2 to Warehouse 2
X_{P2-W3} = number of units to be shipped from Plant 2 to Warehouse 3

TABLE 8-2   Unit distribution costs

	To:		
From:	Warehouse 1 (weekly demand = 80)	Warehouse 2 (weekly demand = 130)	Warehouse 3 (weekly demand = 90)
Plant 1 (weekly capacity = 100)	$7	$ 8	$5
Plant 2 (weekly capacity = 200)	9	12	6

Since each variable represents the number of units to be shipped from one of the plants to one of the warehouses, the objective is to assign values to the six decision variables so that the resulting total cost is minimized. Mathematically, the objective function is simply

$$\text{Minimize cost} = \$7X_{P1-W1} + \$8X_{P1-W2} + \$5X_{P1-W3}$$
$$+ \$9X_{P2-W1} + \$12X_{P2-W2} + \$6X_{P2-W3}$$

Two types of constraints affect the distribution plan (values that the variables can assume in the objective function). The first type is formed by the fact that the total number of units scheduled to be shipped *from* any plant, regardless of their destination, must logically equal the production output of that plant:

Weekly shipments *from* a plant = weekly production *of* the plant

or

Plant 1: $\quad\quad\quad X_{P1-W1} + X_{P1-W2} + X_{P1-W3} = 100$

Plant 2: $\quad\quad\quad X_{P2-W1} + X_{P2-W2} + X_{P2-W3} = 200$

Second, the total number of units scheduled for shipment *to* a warehouse, regardless of their origin, must logically equal the demand *at* that warehouse:

Weekly shipments *to* a warehouse = weekly demand *at* the warehouse

or

Warehouse 1: $\quad\quad\quad X_{P1-W1} + X_{P2-W1} = 80$

Warehouse 2: $\quad\quad\quad X_{P1-W2} + X_{P2-W2} = 130$

Warehouse 3: $\quad\quad\quad X_{P1-W3} + X_{P2-W3} = 90$

Again, the objective function and constraints are combined to provide the following mathematical formulation of the problem.

$$\text{Minimize cost} = 7X_{P1-W1} + 8X_{P1-W2} + 15X_{P1-W3}$$
$$+ 9X_{P2-W1} + 12X_{P2-W2} + 6X_{P2-W3}$$

subject to

$$
\begin{aligned}
X_{P1-W1} + X_{P1-W2} + X_{P1-W3} & & = 100 \\
X_{P2-W1} + X_{P2-W2} + X_{P2-W3} & = 200 \\
X_{P1-W1} \qquad\qquad + X_{P2-W1} & = 80 \\
X_{P1-W2} \qquad\qquad + X_{P2-W2} & = 130 \\
X_{P1-W3} \qquad\qquad + X_{P2-W3} & = 90
\end{aligned}
$$

GENERAL SOLUTION METHODS

The easiest type of linear-programming problem to solve is one in which there are only two decision variables or unknowns in the problem. That was the case in the first example, where the production schedule will be made up of some combination of desk and floor lamps. Since there are only two variables, the objective and constraints can be plotted on a simple graph where the X axis is used to represent one of the decision variables and the Y axis the other. In more complicated (and more realistic) problems a general solution algorithm known as the *simplex method* is used. Most linear-programming computer programs are developed by translating the procedure of the simplex method into a programming language.

Ironically, the solution is probably the easiest step for the user of linear programming. This, of course, is because of the availability of a wide variety of computer packages capable of solving large-scale problems quickly and efficiently. With respect to the two example problems, we have already done the hard part, recognizing and formulating the problem. We now look at some of the solution procedures.

Graphical Method

This solution technique does not have many practical applications since it can be used only when there are two decision variables in the problem. However, it is worth illustrating here because it really shows what linear programming is all about. Sometimes a picture really is worth a thousand words.

Returning to the furniture manufacturer, the problem of how many desk lamps D and floor lamps F to schedule for production in the coming week was formulated as

$$\text{Maximize contribution} = \$10D + \$15F$$

subject to

$$
\begin{array}{ll}
15D + 20F \le 4800 & \text{(Wiring)} \\
10D + 20F \le 3600 & \text{(Assembly)} \\
5D + 8F \le 2400 & \text{(Packaging)} \\
F \le 100 & \text{(Demand for floor models)}
\end{array}
$$

A graphical representation of the problem is developed by letting the X axis represent numbers of floor lamps and the Y axis numbers of desk lamps (it could also be done the other way around). Each of the four constraints forms a straight line on the graph. For example, the 4800 minutes available in the Wiring Department will permit wiring 320 desk models (each requires 15 minutes) or 240 floor models (each requires 20 minutes) or some combination of the two.

Figure 8-1 illustrates the format of the graph and the line created by the constraint for the Wiring Department. Any point on the line represents a combination of desk (D) and floor (F) models that can be wired in exactly the 4800 minutes available in that work area. Points below the line represent combinations of desk and floor models that would not require the full 4800 minutes. The shaded area, then, represents the possible combinations of the two models that could be processed by the Wiring Department in the coming week.

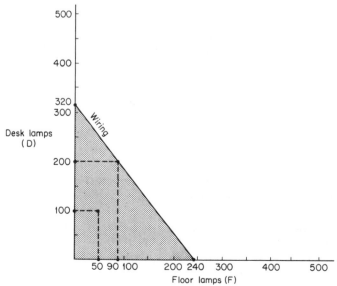

Figure 8-1   Constraint for the Wiring Department.

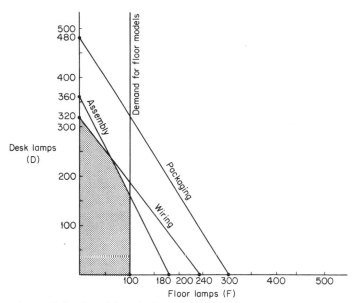

Figure 8-2 Feasible-solution space.

Since the production schedule must satisfy *all* four constraints, the possible combinations of desk and floor models that could be scheduled for production are identified by constructing the other three constraint lines on the graph. This is done in Figure 8-2. The shaded area is the only portion of the graph where all four constraints are satisfied simultaneously. That is, the shaded area represents the model combinations that can be wired, assembled, and packaged within the man-hour allocations and also remain within the limited demand for the floor model. This area is called the *feasible-solution space*. Each point in the space or on the surrounding boundary represents a feasible solution, a value for D and F that meets all the requirements of the problem. One of these points represents the optimal production schedule.

The optimal solution is identified by constructing the objective function on the graph. Since every desk lamp generates $10 in contribution and every floor lamp generates $15, the objective function can also be represented by a straight line. In fact, for any level of total contribution, a line can be constructed on the graph that indicates the combinations of desk and floor models that would have to be produced and sold. As an example, a total contribution of $4500 could be generated by producing and selling 450 desk lamps, or 300 floor lamps, or a number of combinations of the two.

Figure 8-3 shows the objective-function line for a total contribution of $4500. Every point on the line represents a combination of desk and floor lamps that would yield a total contribution of $4500. Also shown in the figure is the shaded area that represents the possible combinations of the two models that can be scheduled for production within the capacity and demand constraints. Since none of the points on the $4500 line lie in the shaded area, there is no feasible production schedule that would yield $4500 in contribution.

Production schedules that would result in $4000 in total contribution are also represented by a line in Figure 8-3. Notice that the $4000 line is below, and parallel to, the $4500 contribution line. The graph illustrates that even $4000 is out of reach. There is no feasible combination of desk and floor lamps that would yield $4000 in total contribution. As the contribution line is lowered in this fashion, it eventually will touch some part of the feasible-solution space. The objective is to find the highest contribution line that touches some part of the shaded area. As the contribution line moves downward, parallel to itself, the optimal solution is identified as the point representing 240 desk lamps and 60 floor lamps. The resulting total contribution is $240(10) + 60(15) = \$3300$. Any other feasible combination of the two models will result in a lower total contribution.

Further Comments on the Graphical Approach

The graphical approach illustrates not only the optimal solution but also many important insights that can be gained about the problem. The

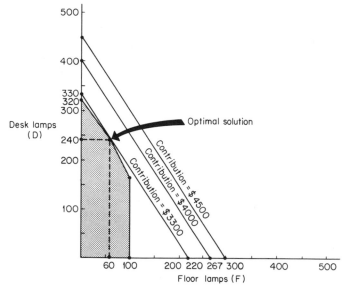

Figure 8-3 Identifying the optimal solution.

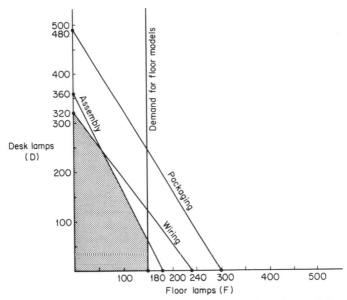

Figure 8-4   Revised feasible-solution space when demand for floor models increases.

shaded area on the graph representing the feasible solutions in Figures 8-2 and 8-3 is determined solely by the constraints in the problem. Figure 8-2 contains a good deal of information. Of the three departments, only Assembly and Wiring are limiting the level of production that can be scheduled. The man-hours available in Packaging are sufficient to engage in a much higher level of production than the 240 desk lamps and 60 floor lamps identified as the optimal solution. The implication is clear: if the man-hour allocations in the departments can somehow be increased (maybe through overtime work), total contribution can be increased if efforts are concentrated in the Assembly and Wiring Departments.

Since the constraints form the feasible-solution space, changes in the demand for floor lamps or the availability of man-hours in the departments could affect the shaded area. If, for example, the demand for floor lamps increased so that as many as 150 could be scheduled for production, the vertical constraint line in Figure 8-2 would move to the right. Figure 8-4 shows what the new feasible-solution space would look like.

Since the shaded area has changed shape, it is possible that the optimal solution has also changed. That is, the highest contribution line that touches some part of the revised shaded area might indicate a new solution. In this particular case, the contribution line would still touch the shaded area at $D= 240$ and $L= 60$, so the solution is not affected. If demand for floor models had fallen so that no more than 50 could be produced, the vertical constraint line would move to the left, forming a new feasible-solution space *and* a new optimal solution. The impact of changes in any of the other constraints can be visualized on the graph in a similar manner. Also, it is possible, using more advanced techniques, to determine the range in which a particular constraint value (such as the demand for floor lamps) could vary without causing a change in the solution.

No matter what form the shaded area takes, it is the objective function or contribution line that identifies the "best" of the feasible solutions. The slope of the line is determined by the unit contribution of each type of lamp. Once the contribution line is positioned on the graph, it is easy to visualize it being moved to coincide with the shaded area, thereby identifying the optimal solution.

If the contribution of either model were to change (due to an increase or decrease in either prices or costs), the slope of the contribution line would also change. This could result in a change in the optimal solution, depending on whether the new contribution line touches the shaded area at a different point. If, for example, the contribution for desk lamps increases from $15 to $25, the contribution line will become steeper. Using the original set of constraints, the relationship between the feasible-solution space and the new contribution line is illustrated in Figure 8-5.

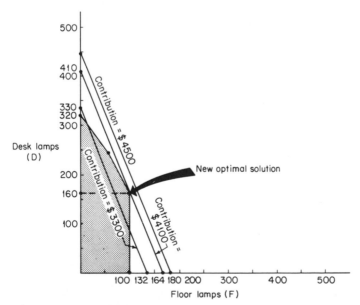

Figure 8-5 Revised solution when contribution for floor models increases.

Since floor models have become more profitable in Figure 8-5, fewer of them have to be produced and sold to generate a given level of contribution. Thus, after the increase in contribution, only 180 floor models will generate $4500 whereas previously 300 were required. Figure 8-5 shows that $4500 in total contribution is still not possible. However, the maximum contribution of $3300 before the increase can easily be surpassed. The fact that the $3300 line runs *through* the shaded area indicates that there are a number of feasible solutions that would yield $3300. It also is clear that the line can move out further and still coincide with at least one feasible solution. The highest contribution line that still touches the shaded area indicates that the optimal solution changes to $D = 160$, $F = 100$ with total contribution = $4100.

If the increase in the contribution generated from floor models had been somewhat smaller, it is possible that the optimal solution would not have changed at all. Again, advanced linear-programming techniques are available to determine the range in which the contribution of each model can vary without causing the solution to change.

An analysis of this sort that is concerned with the effect on the optimal solution of changes in contribution, demand restrictions, and/or man-hour allocations is referred to as *sensitivity analysis*. Often the sensitivity analysis is as important as identifying the optimal solution. Businesses are dynamic in nature, and changing conditions can dictate changes in policy and operating procedures. Sensitivity analysis provides important information on when and how strategies should be changed to accommodate new conditions. Most linear-programming computer programs provide some sensitivity information in addition to the optimal solution. We now examine a more practical solution technique.

Simplex Method

The simplex method is a general algorithm that can be used to solve any linear-programming problem of any size. It consists of arranging the constraint equations in the form of a tableau and systematically generating feasible solutions to the problem. The technique is iterative in nature, where each step in the procedure generates a new tableau that represents a feasible solution. The algorithm terminates when a solution is generated that cannot be improved.

Like most algorithms, the simplex method is a very mechanical procedure that is easily programmed on a computer. The actual mechanics of the algorithm are not important, since virtually all practical applications of linear programming are solved by computer. Instead of getting into the mechanics, we illustrate the nature of the data going into and coming out of a simplex solution done by computer.

Computer Solution

The simplex solution (whether done by computer or by hand) requires that the constraints be expressed as equations rather than inequalities. This is done by incorporating some additional variables into the problem. The computer solution will be illustrated with the production-scheduling problem we have just solved graphically. The constraint for the Wiring Department was:

$$15D + 20F \leq 4800$$

To express the constraint as an equality, a *slack variable* is added to the left side:

$$15D + 20F + S_1 = 4800$$

If the number of desk and floor models scheduled for production does not require the full 4800 minutes, S_1 will take up the slack, or difference. For example, if $D = 100$ and $F = 100$, then S_1 must equal 1300. The slack variable S_1, then, represents unused capacity in the Wiring Department. In a similar manner, the other three constraint inequalities are converted into equalities by adding a slack variable to each

$$10D + 20F + S_2 = 3600 \qquad \text{(Assembly)}$$
$$5D + 8F + S_3 = 2400 \qquad \text{(Packaging)}$$
$$F + S_4 = 100 \qquad \text{(Demand for floor models)}$$

where $S_2 =$ unused capacity in Assembly Department
$S_3 =$ unused capacity in Packaging Department
$S_4 =$ additional floor models that could be sold

In situations where a constraint is of the greater-than-or-equal-to variety, the equality is formed by *subtracting* the additional variable from the left side. In that case, the additional variable is known as a *surplus variable*. In most cases, the slack and/or surplus variables are introduced automatically by the computer, but it is necessary for users to understand how it is done so that they can interpret the resulting computer printout.

Table 8-3 shows the form in which the data are stored in the computer and referred to in the printout. The four constraint equations are shown with all the variables lined up neatly in columns. The zero coefficients indicate that the variable is not present in that particular equation. The data are set up so that each column contains one specific variable in the problem. The format is important because this is how the computer refers to the variables in the printout. Thus, the variable in column 1 *(D)* will be

TABLE 8-3 *Form of the constraint equations*

Variable 1 = D		Variable 2 = F		Variable 3 = S_1		Variable 4 = S_2		Variable 5 = S_3		Variable 6 = S_4			
$15D$	+	$20F$	+	$1S_1$	+	$0S_2$	+	$0S_3$	+	$0S_4$	=	4800
$10D$	+	$20F$	+	$0S_1$	+	$1S_2$	+	$0S_3$	+	$0S_4$	=	3600
$5D$	+	$8F$	+	$0S_1$	+	$0S_2$	+	$1S_3$	+	$0S_4$	=	2400
$0D$	+	$1F$	+	$0S_1$	+	$0S_2$	+	$0S_3$	+	$1S_4$	=	100

referred to by the computer as variable 1; the variable in column 2 *(F)* as variable 2; and so on.

Figure 8-6 presents the computer printout. At the top of the printout, the computer identifies variables 1 and 2 as the decision variables and variables 3 through 6 as the slack variables added to form the equations. The series of numbers that follow are the coefficients in the four constraint equations as they are arranged in Table 8-3. The last row of numbers is used by the computer to generate and evaluate alternative solutions.

Three alternative solutions are evaluated by the program before the optimal solution is identified. In listing the solution at each iteration, the computer indicates only those variables which have nonzero values. For the first solution, variables 1 and 2 (desk and floor lamps, respectively) are zero and the four slack variables take on the values shown. That solution is equivalent to scheduling nothing for production at all and having all the resources unused. Looking back at Figure 8-2, we see that the first solution is represented by the origin on the graph. The simplex algorithm typically starts with a ridiculous (but feasible) solution and systematically moves to the optimal.

The second solution is to produce 100 floor lamps and no desk lamps, which would leave 2800 unused minutes in Wiring, 1600 in Assembly, and 1600 in Packaging. The third solution calls for the production of 160 desk lamps and 100 floor lamps, which would require all the man-hours allocated in Assembly but leave some unused time in Wiring and Packaging. Finally, the solution that maximizes total contribution is identified as $D = 240$ and $L = 60$, which consumes all the time in Wiring and Assembly (variables 3 and 4 are zero) and leaves 720 unused minutes in Packaging. Also, the optimal solution produces 40 fewer floor lamps than the market would absorb (variable 6 = 40). Again looking back at Figure 8-2, we see that the solutions generated by the computer are all extreme points along the border of the feasible-solution area.

```
  * LINEAR PROGRAMMING *


  *******************************************************************

       YOUR VARIABLES 1     THROUGH 2
       SLACK VARIABLES 3     THROUGH 6
  TABLEAU AFTER 0     ITERATIONS
     15     20     1     0     0     0     4800

     10     20     0     1     0     0     3600

      5      8     0     0     1     0     2400

      0      1     0     0     0     1     100

    -10    -15     0     0     0     0     0



       BASIS BEFORE ITERATION 1
       VARIABLE       VALUE
       3              4800
       4              3600
       5              2400
       6              100
       BASIS BEFORE ITERATION 2
       VARIABLE       VALUE
       3              2800
       4              1600
       5              1600
       2              100
       BASIS BEFORE ITERATION 3
       VARIABLE       VALUE
       3              400
       1              160
       5              800
       2              100
       ANSWERS:
       VARIABLE       VALUE
       6              40
       1              240
       5              720
       2              60
       DUAL VARIABLES:
       COLUMN         VALUE
       3              .5
       4              .25
       5              0
       6              0
       OBJECTIVE FUNCTION VALUE = 3300
       IN 3     ITERATIONS
```

Figure 8-6 Computer printout.

The dual variables listed below the optimal solution provide some important sensitivity information. The values for the dual variables indicate how much total contribution could be increased if more man-hours were available in each of the departments or if the demand restriction for floor models was increased. For example, the variables in columns 3 and 4 refer to unused time in Wiring and Assembly, respectively. Since in the optimal solution the man-hour allocations in both departments were completely exhausted, production output and total contribution could be increased if additional time could be obtained. The dual variables indicate that an additional minute in Wiring would permit a production schedule that increases total contribution by $.50 more than the current optimal level of $3300. An additional minute in Assembly is worth $.25. Since Packaging already has unused time in the optimal solution, increasing the man-hour allocation there is worth nothing by itself. Similarly, the optimal number of floor models is only 60, so increasing the maximum production level above the present 100 is also worth nothing. The values for the dual variables are often called *shadow prices,* and in this problem they provide valuable information in deciding if and where overtime work should be scheduled.

Transportation Method

As mentioned earlier, any linear-programming problem can be solved by the simplex method just illustrated. However, certain types of linear-programming problems can be solved more efficiently by another method. Specifically, the general class of problems that we have referred to as transportation problems have their own solution technique. The second formulation problem illustrated in this chapter is an example of a transportation problem. These types of problems are easy to recognize since they usually involve finding the most economical way to distribute something from a set of origins to a set of destinations.

In the earlier example the problem was to distribute the production output of a system of two plants to meet demand at three regional warehouses in such a way that total distribution costs are minimized. The mathematical formulation resulted in an objective-function equation and a series of five constraint equations. In that form, the problem could be entered into a standard linear-programming computer program and solved by the simplex method. However, the problem could also be formulated and solved in a transportation matrix that is very easy to visualize and work with. Table 8-4 illustrates the format of the transportation matrix.

TABLE 8-4   Format of the transportation matrix

		To:			
		Warehouse 1	Warehouse 2	Warehouse 3	
From:	Plant 1	7	8	5	100
	Plant 2	9	12	6	200
		80	130	90	

TABLE 8-5   Optimal distribution schedule

	Warehouse 1	Warehouse 2	Warehouse 3	
Plant 1	7 0	8 100	5 0	100
Plant 2	9 80	12 30	6 90	200
	80	130	90	

Total cost = 0(7) + 100(8) + 0(5) + 80(9) + 30(12) + 90(6)
= $2420

Each square, or cell, in the matrix represents a possible distribution channel, with the unit cost indicated in the upper right-hand corner. The six decision variables in the problem, then, are represented by the cells in the matrix. Any way that the cells can be filled with numbers so that the row and column totals come out right will represent a feasible solution to the problem. For example, the first row represents shipments from Plant 1, and its total must equal 100. The first column represents shipments to Warehouse 1, and its total must equal 80. There are many different ways that the matrix can be filled out. The problem, of course, is to fill it out so that the resulting total cost is minimized.

The solution to the transportation matrix need not be approached by trial and error. The transportation method is a specific algorithm that results in the optimal way to fill out the matrix. Like the simplex method, it consists of a series of step-by-step procedures that are easily programmed on a computer. Most computer installations can also apply this solution technique. The advantage of the transportation method is its efficiency in arriving at the optimal solution to this type of problem and the ease with which the problem can be formulated and conceptualized. The optimal distribution schedule for the example problem is shown in Table 8-5.

CONCLUSION

Linear-programming techniques are very useful in certain types of decision-making situations. With recent developments in computer technology, solutions to large-scale problems can be developed quickly and efficiently. Still, many potential uses of linear programming probably are not taken advantage of because the decision maker simply is not aware of the technique or has not developed a working level of skill in recognizing linear-programming problems. Problem recognition and formulation are unquestionably the key elements in the successful use of linear programming. The sensitivity information produced by most computer solution methods is also an important part of the analysis since it identifies when changes in strategies or operating procedures are appropriate. Finally, implementation of the solution should be done with careful consideration of factors that might be important but were not explicitly considered in the linear-programming model.

CHAPTER 9

Critical-Path Scheduling

At one time or another most managers will be involved in some aspect of a large-scale project. Planning an addition to a plant, installing and debugging a computer system, introducing a new product, and developing and implementing a new wage-incentive system are all examples of projects that require coordination and communication between many different individuals or departments. Very often timing is an important consideration in determining the success of such projects. *Critical-path* scheduling refers to a number of techniques that facilitate the planning, scheduling, and control of large-scale projects.

Critical-path techniques are often referred to as network models, since they depict the various elements of the project in the form of a schematic diagram. The diagram or network illustrates what has to be done to complete the project, the relationships between the various component activities, and an estimate of how long each individual activity should take. From the network it is possible to identify those activities which seem critical to the timely completion of the project.

There are many advantages to this type of approach. The use of a critical-path technique lends structure to what is often a very complex situation. Sometimes there are few if any precedents to rely on in planning the project. In most cases, information must come from a number of sources, both internal and external, in order to estimate the time required to complete the project. The time estimate is often extremely important, especially when the project involves such things as introducing a new product or constructing a building, where penalty costs are incurred for delays. The network representation facilitates planning by providing a framework that illustrates how each segment of the project fits into the overall picture.

The network also provides a means of monitoring the project as it progresses. It is possible to compare the actual progress made at any time with the expected state of completion from the network diagram. In this fashion, the impact of delays in some of the component activities can be estimated and the appropriate actions taken *before* the entire project overruns its scheduled completion.

Finally, most critical-path techniques assist the planner in making best use of the resources that are available. Typically, a limited amount of money, equipment, and manpower is available to accomplish the project. It usually works out that some individual activities can be delayed with little or no consequence to the overall completion schedule, whereas a delay in another activity will affect completion of the entire project. By identifying the individual activities that potentially can delay the entire project the network focuses attention to those aspects of the project where resources should be concentrated.

TYPES OF CRITICAL-PATH TECHNIQUES

The two original and most widely used critical-path techniques are the Performance Evaluation and Review Technique (PERT) and the critical-path method (CPM). PERT was developed in the late 1950s by the U.S. Navy and a management consulting firm, for use in planning and controlling the Polaris missile project. CPM was developed by the Du Pont Company for internal control of maintenance and engineering projects. Both techniques are now well known and have been used successfully by many companies. In fact, some government agencies now require their subcontractors to submit a PERT or CPM schedule before awarding a contract.

PERT and CPM were developed at about the same time and are similar in concept. Both are based on a network representation of the activities in the project. The difference between the two lies in how the network is constructed. That is, the two techniques differ slightly in how the component activities of the project are represented schematically. Another difference is that CPM allows one to incorporate cost estimates as well as time estimates for each individual activity. A PERT analysis is concerned mostly with time. Also, in PERT time estimates for the component activities are collected in such a way that a statistical analysis can be performed to determine the probability of finishing the project within a certain time. In CPM a single estimate of the project length is derived, and no probabilities or confidence intervals can be computed.

Aside from PERT and CPM, a number of other critical-path scheduling techniques have been developed. They come with a wide variety of catchy names, but their basic purposes are the same. Although they may vary slightly in their intended application, all critical-path techniques have two things in common: a network to represent the project and the ability to identify the critical activities in completing the project on schedule. Since it would be confusing to try to illustrate the many different critical-path techniques, we focus our attention on PERT in the remainder of this chapter. The PERT technique is widely used in government and industry, and anyone who has a conceptual understanding of PERT will have no difficulty in learning how to use any critical-path technique.

WHEN IS PERT APPROPRIATE?

One of the advantages of PERT (or any critical-path technique for that matter) is that it can be applied to a wide variety of situations. There are virtually no mathematical assumptions made that limit the practical applications of the technique. PERT can be applied with equal effectiveness by a large industrial corporation planning the construction of a new plant and a scientist coordinating the various aspects of a research project. The fact that the range of applications is virtually unlimited helps explain why critical-path techniques have caught on so quickly in management circles.

In general, the use of PERT requires three things. First, the project must be broken down into separate components or activities. Each activity represents some part of the project that must be completed. There are no set rules on how to define the activities. In some cases, a very broad identification of what constitutes an activity will be good enough. For example, in a construction project an activity might be defined as "purchase materials." In other cases, more detailed information might be needed and the activities defined more narrowly, such as "determine possible vendors," "evaluate and select vendor(s)," and "issue purchase orders."

The appropriate way to define the activities or components of the project is determined by the degree of control needed. If the project requires contributions from different departments or functional units in an organization, better control could probably be achieved by defining the activities so that the contributions of the different departments are viewed as separate elements. The PERT technique will work with any description of the activities.

Second, the order in which the activities are to be performed must be determined. Some activities can be performed simultaneously, while others can begin only when some preceding activity is finished. In the construction example, the excavation of the building site and the purchase of the necessary materials are two activities that probably can be done simultaneously. On the other hand, building the frame of the structure and constructing the roof are two activities that must be done sequentially. These relationships are important because a large number of sequential activities in a project means that a delay in one affects them all. The PERT network is a schematic way of illustrating the relationships between the activities in the project.

Finally, each individual activity must be accompanied by an estimate of how long it will take. Each time estimate comes from an individual or group that has expertise in that particular activity. While the project may be a one-time venture, there are usually people in an organization familiar with the component activities that make up the project. The purchasing manager can probably provide a meaningful estimate for how long it will take to get the materials, and the foreman at the site can probably make a good guess about how long the excavation will take.

In PERT there are actually three time estimates given for each activity: a most likely estimate, a pessimistic estimate, and an optimistic estimate. The three estimates are combined to get an expected time for each activity in the project, and the range in the three estimates is often used as a gauge for the level of confidence that should be placed in the results of the PERT analysis. Before illustrating the procedure with a specific example, we give a more precise definition to some of the terms we have been using, as well as those needed later.

PERT Terminology

Terminology is important in understanding and communicating many management science techniques. The following terms have a specific meaning in PERT:

Activity   A subunit or component of the project. Breaking the project down into activities forms the basis for the PERT analysis.

Event   A milestone or checkpoint in the completion of the project. For example, the completion of some preceding activity is an event that must occur before work on the following activities can begin.

Network   A schematic diagram that illustrates the activities in the project and the order in which they must be performed. In PERT, arrows are used to represent activities and circles or nodes to represent events. The series of connecting arrows and nodes illustrates the relationships between the component parts of the project.

Path   A sequence of activities that is represented by connecting arrows through the entire network. Each path represents a series of activities that must be done in sequential order.

Critical Path   The path that requires the longest time to complete. The critical path provides the best estimate for the length of the project. Activities on the critical path are those with the potential to delay the entire project completion.

Slack   The amount of delay that can be tolerated in reaching an event without delaying the whole project. Events not on the critical path usually have slack associated with them. Slack can also be expressed in terms of the allowable delay in the completion of each individual activity in the project.

AN ILLUSTRATIVE EXAMPLE

A company has decided to conduct a market survey to obtain information needed to evaluate and revise its marketing programs. The survey is to be conducted by trained interviewers, who will administer a questionnaire to members of preselected households. The survey information is needed before the company can proceed with developing its marketing strategy. Management has decided to develop a PERT schedule for the project.

Table 9-1 illustrates how the activities in the project are defined, the

TABLE 9-1   *Marketing survey project*

Activity	Code	Immediately preceding activity	Time estimate, days		
			Optimistic	Most likely	Pessimistic
Develop questions to ask in survey	A	—	3	5	8
Recruit interviewers to administer questionnaire	B	—	5	7	12
Conduct training session for interviewers	C	B	1	1	1
Analyze and validate questions to be included in the survey	D	A	2	3	5
Have questionnaire printed	E	D	3	7	10
Develop sampling procedure for selecting households	F	—	2	4	6
Administer the questionnaire	G	C,E,F	8	10	18
Tabulate results and run statistical tests	H	G	5	7	10
Summarize findings and their implications to top management	I	H	3	5	10

precedence relationships between the activities, and their individual time estimates. Each activity is given a code for easy identification. The column entitled "Immediately preceding activity" shows whether there is a sequential relationship between one or more activities. For example, the table illustrates that the training session for the interviewers (activity C) cannot logically be conducted until the interviewers have been recruited (activity B). Thus, a delay in recruiting interviewers for the survey will affect when the training session can be held. In addition, the actual administration of the questionnaire cannot begin until the training session has been completed (activity C), the questionnaires have been printed (activity E), and the sample households selected (activity F). Again, a delay in any one of these activities affects the entire sequence. Activities A, B, and F have no preceding activity, which implies that work on them can begin as soon as the project gets under way.

The three times estimates for each activity suggest a distribution of possible completion times. As mentioned earlier, the time estimates should come from people with experience or expertise in the activity. The optimistic estimate is the time that would be required for the activity if everything goes as well as it possibly could. More precisely, the optimistic estimate should be set so that the probability of being able to complete the activity in something less than the optimistic time is virtually zero. On the other hand, the pessimistic time is an estimate of how long the activity would take if just about everything that could go wrong actually does. The probability that the activity will require more than the pessimistic time should be virtually zero. Finally, the most likely time is just what its name implies. If the activity were repeated many times, one would expect the most frequently observed time to be the most likely estimate.

In Table 9-1 the time estimates indicate that it would be virtually impossible to develop the questions for the survey in less than 3 days. On the other hand, under no circumstances should it take more than 8 days. More than likely, neither extremely favorable nor extremely unfavorable circumstances will prevail, and the questions should be developed in 5 days. The data also indicate that the training session for the interviewers will require only 1 day. The progress made in recruiting the interviewers (activity B) will affect when the training session can begin, but its duration will not exceed 1 day. This method of deriving three time estimates provides a useful way for the estimator to express uncertainty about the activity times. Further, a wide variation between the optimistic and pessimistic times for an activity should not be interpreted as ignorance on the part of the estimator. It implies instead that a number of factors (weather conditions, equipment failure, material shortages, and so forth) could significantly affect the activity's completion time.

Computing Expected Times for the Activities

The usual PERT procedure is to combine the three time estimates into an expected completion time for each activity. The method is very simple. The following formula is used to compute the expected completion time t_e of each activity in the project:

$$t_e = \frac{o + 4m + p}{6} \qquad (9\text{-}1)$$

where o is the optimistic estimate, m is the most likely estimate, and p is the pessimistic estimate.

The formula is really nothing more than a weighted average, with the highest weight being placed, of course, on the most likely estimate. If unfavorable conditions would cause a significant delay in the activity, the expected time would reflect this by taking on a value that is greater than the most likely estimate. Conversely, if extremely favorable conditions enable the activity to be completed very quickly, the expected time would be less than the most likely estimate, reflecting the chances of a quick completion.

While the calculation is simple and straightforward, the underlying statistical theory is a bit more complicated than the formula implies. The time that is going to be required to complete an activity is really an unknown or random variable that can be described by some probability distribution. In PERT, a theoretical probability distribution known as the beta distribution is used to represent the individual activity completion times. The optimistic and pessimistic times are really estimates of the endpoints (or range) of the distribution, while the most likely time is an estimate of the mode of the distribution. In computing t_e for each activity, we are really using the three parameter estimates to estimate the mean or expected value of the beta distribution. The beta distribution is not really important to us here, but the concepts of parameter estimation and theoretical probability distributions should be familiar from Chapters 4 to 6. In Table 9-2 the expected times for the activities in the example problem are computed.

Constructing the PERT Network

A PERT network for the market survey is constructed by developing a schematic diagram in which each of the nine activities is represented by arrows. The arrows begin and end at events (represented by nodes) which illustrate the precedence restrictions in performing the activities. The PERT network is a graphical way of summarizing the data in Tables 9-1 and 9-2. Figure 9-1 illustrates the PERT network.

Beneath each arrow in the network is the activity that it represents and above the arrow is the activity's expected completion time (from Table 9-2). Each node is numbered for easy identification and represents an event

TABLE 9-2   Expected activity times

Activity code	Calculation	Expected time t_e, days
A	$\dfrac{3 + 4(5) + 8}{6}$	5.17
B	$\dfrac{5 + 4(7) + 12}{6}$	7.50
C	$\dfrac{1 + 4(1) + 1}{6}$	1.00
D	$\dfrac{2 + 4(3) + 5}{6}$	3.17
E	$\dfrac{3 + 4(7) + 10}{6}$	6.83
F	$\dfrac{2 + 4(4) + 6}{6}$	4.00
G	$\dfrac{8 + 4(10) + 18}{6}$	11.00
H	$\dfrac{5 + 4(7) + 10}{6}$	7.17
I	$\dfrac{3 + 4(5) + 10}{6}$	5.50

or milestone in the completion of the project. For example, event 1 represents the beginning of the market survey project. At that time, three separate activities can begin simultaneously: developing the questions (A), recruiting the interviewers (B), and developing the sample (F). The three activities are represented by arrows that start at event 1.

Event 2 represents the completion of activity A. That is, event 2 represents the point when the questions to be asked in the survey have been developed. At that point, efforts to analyze and validate the questionnaire (activity D) can begin. In a similar fashion, every node in the network has a specific meaning in terms of the completion stage of the project, and the following activity arrow represents the next step in the project.

The dotted arrows in the network in Figure 9-1 are not activities as defined in the project but "dummy activities" necessary to make the network reflect the precedence relationships accurately. The need for the dummy activities becomes obvious when one examines the network. Event 7 represents the completion of the sampling procedures (activity F), while event 6 the completion of the printing (activity E), and event 4 the completion of the training session (activity C). Naturally, all three activities must be completed before the questionnaire can be administered (activity G). If the arrow representing activity G were connected directly to event 7, it would indicate that the questionnaire could be administered as soon as the sample was developed, regardless of the progress made in printing the questionnaire and training the interviewers. Similarly, if activity arrow G were connected directly to either event 6 or event 4, it would indicate that only the one activity had to be completed before the questionnaire could be administered. To reflect the true situation, event 8 is used to represent the time when *all three* preceding activities are completed.

In general, there are a few simple rules to follow in constructing a PERT network. The shape of the network, of course, is determined by the precedence relationships between the activities. The network for this example was drawn from the precedence data in Table 9-1. If possible, it is desirable to construct the network so that the activity arrows do not cross. Crisscrossing arrows make it difficult to read the network and visualize the relationships between the activities. Also, there is a definite flow in the network (from left to right), and the activity arrows should be constructed to reflect the direction of flow.

The direction of flow should also be reflected in the way the events (nodes) are numbered for identification. It makes no difference what numbers are used to identify the events as long as the identification numbers appear in increasing order as one moves from left to right through the network. This is particularly important when the computer is used in the PERT analysis, since most computer programs are designed to determine direction of flow in the network from the event numbers.

Finally, the network should have one beginning and one ending node. That is, there should be a single event that indicates the project is under way and a single event that indicates the project is completed. What happens in between, of course, is determined by the activities that have to be

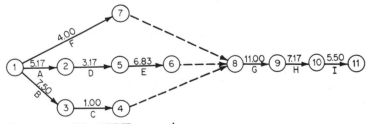

Figure 9-1   The PERT network.

performed and their relationships. It is also important to understand that every arrow and every node has a specific meaning or provides some specific information. We now illustrate how to develop and use the information present in the PERT network.

Identifying the Critical Path

By looking at the PERT network in Figure 9-1 it is easy to identify sequentially related activities. In fact, each path through the network represents a sequence of activities that must be done in that particular order. For example, one path through the network is represented by events 1-7-8-9-10-11. That path simply represents the fact that the sample must be determined before the questionnaire can be administered, tabulated, and summarized.

Another path is developed by events 1-2-5-6-8-9-10-11 and represents the fact that the questionnaire must be produced (questions developed, validated, and printed) before it can be administered and analyzed. The only other path through the network is events 1-3-4-8-9-10-11, representing the fact that there must be interviewers (recruiting and training) before there can be interviews. By summing the expected activity times along each path, the expected time for each sequence of activities is computed. Table 9-3 summarizes the results and identifies the critical path.

The longest path is critical because it represents the sequence of activities that is expected to require the most time. All the activities depicted in the network must be completed before the project is finished, but it would be unreasonable to expect to finish the entire project in a shorter time than the sequence of activities expected to take the most time. Thus, the length of the critical path provides the best estimate for the duration of the project. In this case, it is expected that 38.84 days will elapse from the beginning of the market survey project before the report containing the necessary marketing data is ready to submit to top management.

Aside from the time estimate, the critical path also identifies the activities that are critical to the 38.84-day completion. Since the critical activities (activities on the critical path) are already the most time-consuming, a delay in any one of them makes the critical path even longer and jeopardizes the expected completion in 38.84 days. Of the three general phases of work that must be completed before the questionnaire can be admin-

TABLE 9-3 **Paths through the network**

Path	Total expected time, days	
1-7-8-9-10-11	27.67	
1-2-5-6-8-9-10-11	38.84	critical path
1-3-4-8-9-10-11	32.17	

istered (questionnaire, sample, interviewers), it is more critical to the completion of the project that work in the development of the questionnaire go on according to schedule. In fact, a delay in anything having to do with developing, validating, printing, administering, and analyzing the questionnaire will cause an expected delay in the entire market survey. On the other hand, noncritical activities such as developing the sample (F) and recruiting and training the interviewers (B and C) could conceivably be delayed without affecting the 38.84-day estimate at all. The amount of delay that can be tolerated in the noncritical activities is referred to as *slack*.

Computing the Slack

To compute the slack times we first determine when each of the 11 events would be expected to occur if the project proceeds on schedule. We have already determined that if things go as expected, the project will be finished in 38.84 days. The time when each event is expected to occur T_E is easily computed from the network. Starting at time zero for the beginning of the project (event 1), the questions should be developed (event 2) after 5.17 days, the recruiting completed (event 3) after 7.5 days, and the sampling procedures determined (event 7) after 4 days.

Proceeding further, if things go as expected, the validation of the questions should be finished (event 5) after $5.17 + 3.17 = 8.34$ days and the printed questionnaire should be ready (event 6) after $8.34 + 6.83 = 15.17$ days. For the interviewers, the training session is expected to be completed (event 4) after the project has been under way for $7.50 + 1.00 = 8.50$ days. Event 8, representing the time that the questionnaire can be administered, cannot occur until events 7, 6, and 4 have taken place. Thus, the administering of the questionnaire could not be expected to begin until 15.17 days have elapsed. The T_E values for the remainder of the events are calculated by proceeding through the network in identical fashion.

To determine the slack times, we must also compute the *latest* time that each event can occur T_L in order to finish the project in the 38.84-day estimate. The slack for each event is defined as the difference between the latest time T_L the event can occur and the expected time T_E just illustrated. The T_L values are easily calculated by starting at the *last* event in the network and working backward.

To illustrate the T_L calculations, it is obvious that the latest time that event 11 can occur is 38.84 if the project is to be completed on schedule. That means that the results of the survey must be tabulated (event 10) after $38.84 - 5.50 = 33.34$ days if the completion schedule is to be met. Working back farther, all the questionnaires will have to be administered (event 9) after $33.34 - 7.17 = 26.17$ days. This means that the questionnaire must be ready to go (event 8) after $26.17 - 11 = 15.17$ days. It also

TABLE 9-4   Computing slack times

Event	Represents completion of	Expected to occur (T_E)	Must occur (T_L)	Slack ($T_L - T_E$)
1	(Beginning)	0	0	0
2	A	5.17	5.17	0
3	B	7.50	14.17	6.67
4	C	8.50	15.17	6.67
5	D	8.34	8.34	0
6	E	15.17	15.17	0
7	F	4.00	15.17	11.17
8	F,E,C	15.17	15.17	0
9	G	26.17	26.17	0
10	H	33.34	33.34	0
11	I	38.84	38.84	0

means that the latest possible time to have developed the sampling procedures (event 7), printed the questionnaire (event 6), and trained the interviewers (event 4) is 15.17 days, since all must be accomplished before the questionnaire can be administered. The T_L values for the remaining events are computed by simply working back through the network in this fashion to event 1. Table 9-4 summarizes the T_E and T_L calculations and shows the resulting slack for each event in the project.

The table shows that all events representing the completion of critical activities have zero slack. That is precisely why those activities are critical: a delay in the completion of any of them causes an expected delay in the entire project. The calculations also show how much delay can occur in completing the noncritical activities (B, C, F) without affecting the 38.84-day schedule for the entire market survey. For example, the recruiting of the interviewers (activity B) is expected to be completed after 7.5 days but does not have to be completed until 14.17 days have elapsed. The difference of 6.67 days represents the allowable delay. The table also reveals that completion of the training session (activity C) could be delayed as much as 6.67 days and completion of the sampling procedures (activity F) could run as much as 11.17 days behind schedule without affecting the project. Each slack time represents the allowable delay in completing the activity *if all other activities are completed on schedule*. If actual delays occurred in two or more activities *simultaneously*, it would be necessary to recalculate the T_E and T_L values for the remaining activities in the project.

Implications for Planning and Scheduling

The data developed from the PERT analysis in the preceding sections provide important information in planning the market survey project. It is not at all obvious which activities are critical until the various aspects of the project are depicted in the PERT network. From the critical path comes a meaningful estimate for the length of the project and a knowledge of which activities must be monitored closely. Sometimes this information can be used to expedite the completion of the project.

Suppose, for example, that management needed the market survey information sooner than the 38.84 days estimated from the PERT analysis. What can be done to speed up the expected completion of the project? Which activities should the project manager try to expedite? Partially, at least, the answers should be obvious. The expected completion of the project can be shortened by somehow reducing the length of the critical path. This means that the activities on the critical path are the ones that must be examined for possible savings in time. Expediting an activity not on the critical path would yield no benefits at all in shortening the expected length of the market survey project.

In general, expediting an activity generates additional costs for the extra manpower, equipment, or other resources needed to speed up the expected completion of the activity. The additional costs will be justified if they are outweighed by the benefits resulting from a speedier project completion. What the PERT network provides is a framework for determining the impact on the entire project of expediting certain activities. This allows the project manager to compare the costs and benefits in order to determine whether it is economically feasible to attempt to shorten the expected project completion. This type of analysis is often referred to as *crashing* the project.

In the example problem, the administering of the questionnaire (activity G) could probably be done in less than the expected 11 days if more interviewers were hired. Since this is a critical activity, a reduction in its expected time would result in a reduction in the expected time for the entire market survey project. However, in addition to the extra cost of hiring more interviewers, the time needed for recruiting interviewers (activity B) might very well *increase*, since more people would have to be recruited. But we have already determined that the recruiting could be delayed by nearly 7 days without affecting the completion schedule. Thus, one alternative for crashing the project would be to hire additional interviewers, which would shorten activity G and hasten project completion even though it might lengthen activity B. Again, it would be a management decision whether the benefits of a shorter completion outweigh the additional costs.

The concept of slack time for the noncritical activities can sometimes be used to find a better distribution of existing resources at no additional cost. Looking at the PERT network in Figure 9-1, we see that work on the sampling procedures (activity F) and the questions for the survey (activity A) can begin right away. However, the slack calculations show that work on the sampling procedures can be delayed by 11.17 days with no effect on the project completion. This suggests the possibility of shifting resources from activity F to activity A. For instance, some personnel scheduled to begin working on the sampling procedures might be better used in helping develop the survey questions. The likely result would be to shorten activity A at the expense of activity F. As long as the delay in activity F does not exceed 11.17 days, the expected completion of the entire project is shortened at no additional cost. This reallocation from noncritical to critical activities is often helpful in cases where the resources are transferrable. Of course, money (budget allocation) is one resource that is almost always transferrable.

This type of analysis is often a significant part of the PERT study, especially when time is an important variable in the success of the project. Sometimes the various changes or reallocations will cause the critical path to change. This would happen for example, if activities on the critical path were expedited to the extent that they no longer represent the longest path through the network. Of course, the PERT network illustrates how *all* paths are changed by crashing certain activities, and the knowledgeable manager can use this information to determine the extent to which it would be worthwhile to try to expedite critical activities.

Confidence in the PERT Analysis

It was mentioned briefly that a unique feature of PERT is its ability to make some statement about the confidence that should be placed in the estimated length of the project. Since three time estimates are given for each activity, it is possible to look at the difference between the pessimistic estimate and the optimistic estimate for each activity as a measure of the uncertainty that exists concerning how long that activity will take. As the uncertainty concerning the completion of activities on the critical path *increases,* the confidence that should be placed in the project completion estimate *decreases.* In fact, it is possible to use the pessimistic and optimistic times to estimate the variance of the probability distribution that describes each activity and use that information to establish a confidence interval around the estimate of project completion. We do not illustrate that analysis here but emphasize the point that confidence in the project completion schedule is directly related to the confidence exhibited by the people who provided time estimates in the project.

CONCLUSION

Critical-path techniques provide a framework for effectively planning and controlling large-scale projects. Many of the calculations illustrated in this chapter can be done automatically by standard computer packages. However, computer programs are designed for computational convenience, and the user must have a conceptual understanding of what the computer is doing in order to make effective use of the information that results. The concepts and terminology presented here in the PERT illustration are typical of most critical-path scheduling techniques.

CHAPTER 10

Basic Inventory Models

Inventories play a major role in the operation of many businesses. In manufacturing, inventories of raw materials allow companies to operate somewhat independently of their sources of supply. Day-to-day operations are not critically dependent on deliveries from suppliers since stocks of the necessary materials are maintained and used as needed. Manufacturers also use inventories of finished products to level, or "smooth out," production schedules. By producing to inventory during slack periods and drawing from that inventory during peak sales periods the efficiencies of continuous production can be attained even for products whose demand is highly seasonal.

Retailers and wholesalers maintain inventories for much the same reasons. Inventories of products serve to separate, or "decouple," the retailer and wholesaler from their sources of supply. The result is increased flexibility in operations and better service to customers. As a result of the inventories maintained by retailers, an overwhelming majority of the everyday items required by consumers can be purchased on the spot.

In addition to raw materials and finished products, many companies also maintain sizable inventories of work-in-process as well as office supplies, business forms, and general operating supplies. For some companies, the total number of different items maintained in inventory runs into the hundreds. Often, the amount of money invested in inventories accounts for a significant portion of the company's total assets. In some companies, the figure is as high as 40 or 50 percent. In such cases, the financial structure of the company is significantly affected by inventory decisions.

THE COST OF INVENTORIES

In many ways inventory is a necessary evil. There is no way that rational systems for production and distribution could be developed without the use of inventories. Yet, inventory in and of itself generates no value. In fact, an inventory really represents an investment in something that, in a financial sense, yields no return. At the very least, money that is tied up in inventory could be in the bank earning interest. Further, maintenance of inventories also requires warehouse space, materials-handling equipment, and special record-keeping systems and administrative procedures.

Most managers realize that inventory decisions can have a major impact on a company's operating costs. Too little inventory results in frequent orders and costly stock-outs. Too much inventory ties up needed cash, facilities, and personnel unnecessarily. In some cases, it is possible for a company to reduce its overall inventory-related costs without any sacrifice to operating efficiency. This is especially true when decisions regarding inventory levels and order quantities have been made arbitrarily. Arbitrary decisions to order an item quarterly or semiannually may be convenient but may also generate costs that are significantly higher than they need be.

In this chapter we examine some basic analytic inventory models. The analytic models replace arbitrary decision rules with a more systematic approach that explicitly considers the costs involved. In general, the costs created by maintaining an item in inventory fall into three categories: ordering cost, carrying cost, and stock-out cost.

Ordering Cost

Ordering cost refers to the cost associated with replenishing the inventory. For purchased items, this typically includes administrative cost for preparing and processing the purchase order, shipping and handling cost, and labor cost for inspection and quality control procedures. For items made in-house, the ordering cost is more appropriately called *setup* cost. This includes labor and equipment costs associated with starting up a production run for the item.

The ordering cost (or setup cost) is usually regarded as being fixed. That is, the cost associated with replenishing an inventory usually does not vary with the size of the replenishment order. For example, the administrative cost incurred in preparing and processing a purchase order is not affected by the size of the order. It takes the clerk or typist the same amount of time to enter a month's supply on the purchase order as it does to enter a year's supply. In many cases, shipping cost is the same whether the truck or freight car is completely full or only half full. Similarly, many quality control procedures are based on predetermined sample sizes, so that even inspection cost is relatively independent of order size. For example, a random sample of 50 cases of an incoming material may be broken down for inspection, regardless of whether the total shipment contains 1000, 2000, or even 5000 cases. For items manufactured in-house, the tooling cost to set up the production line does not depend on the size of the production run.

Ordering cost for an inventory item is usually expressed in terms of dollars per order (or setup). In the inventory models presented here we shall assume that this cost is completely fixed. This, of course, does not have to be the case in order to develop an inventory model; models can be developed that describe almost any set of conditions or circumstances. The assumption of fixed ordering cost adequately describes many actual situations and allows the formulation of an elementary inventory model.

The fact that there is a cost associated with the ordering process encourages the inventory manager to issue large replenishment orders when inventory levels are depleted. For the forecast need for an item over a given period, large replenishment orders mean fewer cycles through the reordering process and lower inventory ordering cost over the planning period. At the extreme, the entire quantity needed during the period (or even a supply to last several periods) could be ordered at once. Of course, the ordering cost must be considered with all the other relevant costs before the appropriate order quantity can be determined. Intuitively, however, one would expect that the more costly the ordering process the more likely it is that large replenishment orders will be appropriate.

Carrying Cost

Carrying cost refers to the cost associated with maintaining the item in inventory. A major portion of this category is the opportunity cost of tying up funds in inventory. Other costs created by carrying inventory are for insurance, property taxes, spoilage, and obsolescence. Also, there is a cost associated with materials handling, storage space, and the record keeping necessary to keep track of inventory levels. This type of cost is incurred regardless of whether the item is purchased from a supplier or made in-house.

In general, carrying cost is considered to be variable. That is, as the stock of a particular item gets larger and larger, so does the cost associated with maintaining an inventory of the item. This is logical since a larger inventory increases the amount of capital that is tied up, as well as the insurance premiums for protection against fire and theft and the amount of property taxes. Similarly, a larger inventory increases the potential loss if the item becomes obsolete or is subject to spoilage. The cost associated with materials handling and physically taking inventory also would be expected to increase with the level of stock.

The cost to carry an inventory item in stock is usually expressed in terms of dollars per unit per time period. Sometimes the cost figure is estimated as a percentage of the value of the item. If annual carrying cost for a $1 item were estimated at 20 percent, it would cost an estimated $.20 to hold 1 unit of the item in stock for a period of 1 year. While 20 percent is a commonly used figure, the appropriate cost estimate must be based on the specifics of the situation. Highly perishable items or items that require special facilities (such as refrigeration) or special materials-handling equipment might cost as much as 50 percent of their value or more to carry in inventory for a year.

The inventory models presented here will assume that carrying cost is completely variable. In general, this is often an adequate description of reality and results in a straightforward inventory model. More complex situations can also be incorporated into inventory models, but the basic models presented here will serve to illustrate the usefulness of analytic approaches to inventory management.

The fact that there is a cost associated with carrying stock influences the inventory manager to keep low levels of inventory. Frequent orders for small quantities of the item serve to keep the inventory level low and minimize its carrying cost over the planning period. The higher the estimated carrying cost for an item, the more important it is to make small replenishment orders when the stock is depleted and to maintain a relatively low inventory level. The ordering cost and carrying cost influence the decision maker in opposite directions. Ordering cost encourages large order quantities and high inventory levels; carrying cost encourages small order quantities and low inventory levels. An inventory model provides a framework in which these two opposing forces can be traded off to develop the most economical inventory policy.

Stock-Out Cost

Stock-out cost results when an item is needed but its inventory level is completely depleted. In a manufacturing system, a stock-out might cause production delays, idle labor and equipment, and/or an emergency order to the supplier. In a wholesale or retail operation, a stock-out may lead to lost sales and damaged goodwill.

Stock-out cost can be expressed in either of two ways. Sometimes it is appropriate to estimate a fixed cost that is incurred each time a stock-out is encountered. This might be appropriate for a raw-material item used in manufacturing where a stock-out of the item delays production, regardless of the exact number of units short. Once it is determined that therc is a shortage of the item, the production delay has occurred and the cost associated with the delay does not depend on the total shortage. In other cases, stock-out cost is more appropriately estimated as a variable cost with respect to the number of units short. This might be appropriate for an inventory item that is sold directly to consumers. In that situation, failure to meet demand results in lost sales, and the total cost of the stock-out depends on the number of units that were demanded but not available.

While annual ordering cost and annual carrying cost are largely deter-mined by the *quantity* of replenishment orders, annual stock-out cost is determined mostly by the *timing* of the orders. A stock-out of an item is likely to occur between placing a replenishment order and actually receiv-ing it. If the delivery time takes longer than expected or demand during the period is greater than forecast, a stock-out is likely to result. As the consequences (or cost) of being out of stock become greater and greater, replenishment orders can be placed earlier and earlier to increase the probability that the remaining stock will be sufficient to meet demand until the order arrives. Yet the effect of placing replenishment orders earlier is to increase the overall level of stock and the annual cost associ-ated with carrying the inventory. The appropriate point at which to place the replenishment order must be determined by examining both the stock-out cost and carrying cost for the item in question. This is a problem that will be addressed later.

In most cases, it is a difficult task to estimate the ordering, carrying, and stock-out costs for an inventory item. However, without some idea of the magnitude of these costs, inventory decisions become arbitrary. As we shall illustrate with some elementary models, arbitrary decisions can be much more costly than decisions based on relevant factors whose values can be estimated only roughly.

ABC CLASSIFICATION SYSTEM

Before illustrating the development and application of an analytic inven-tory model, it will be worthwhile to examine a useful system for classifying inventory items. It was mentioned earlier that some companies maintain hundreds of different items in inventory. In such cases, it is virtually impossible to maintain strict control over the stock levels and order quan-tities of every item. In fact, it is not even desirable to attempt to do so. For many of the items, the costs of attempting to apply strict inventory con-trols simply outweigh the benefits.

The ABC classification system is a method for identifying the inventory items for which individual control over stock levels and order quantities is likely to yield substantial cost savings. Usually, the most important items to include in an inventory control system are those which require the largest investment. The more expensive items tie up larger amounts of capital and also generate higher costs for insurance, taxes, and potential spoilage or obsolescence. It follows that the consequences of being overstocked are much greater if the item is an expensive one. Also, the more expensive items are typically important parts or subassemblies of the product, and so the consequences of understocking are also severe.

The ABC classification is based on a simple concept that holds true for many inventory systems. Usually, a small percentage of the total number of different items maintained in inventory makes up a very large percentage of the total dollar investment in inventories. These relatively few items are referred to as class A items, and they usually require strict inventory controls. On the other hand, the vast majority of the inventory items, when taken together, usually account for only a small percentage of the total investment in inventories. These items are the class C items, and they usually warrant only very loose control. The class B items are those that fall somewhere between these two extremes, and they usually require moderate control procedures.

Figure 10-1 illustrates graphically the concept underlying the ABC classification system. Although the exact percentages will vary from one system to another, typically as few as 10 to 15 percent of the inventory items accounts for as much as 75 percent of the total inventory value. The mid-

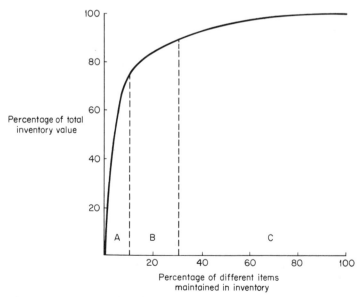

Figure 10-1   ABC classification.

dle category, class B, usually consists of the 15 to 20 percent of the inventory items that accounts for approximately 15 to 20 percent of the total investment in inventories. Finally, the chart illustrates that class C items typically include as much as 70 percent of the inventory items, but account for only about 10 percent of total inventory value.

A major benefit of the ABC classification system is that it lends some degree of order to the situation. It becomes obvious from Figure 10-1 that strict control over a relatively small number of items will go a long way toward minimizing the cost associated with maintaining the entire inventory system. It is not difficult to identify the class A items, and for very large systems computer programs are available to make the classifications automatically.

Most of the analytic approaches to inventory management are concerned with determining the most economical stock level and order quantity for an item that falls in class A. This focus of attention on class A items is not surprising, since these items make up such a large part of the inventory investment. We now illustrate the development and application of an elementary but very important analytic inventory model.

THE ECONOMIC-ORDER-QUANTITY (EOQ) MODEL

Suppose a manufacturer of television sets maintains an inventory of an electronic switch that is used in certain models. The electronic switches are purchased from a supplier at a cost of $10 each. In order to keep the illustration as simple as possible, let us assume that the production line operates 300 days per year and uses 40 of the electronic switches each day (12,000 per year). When the inventory of switches drops to a certain point, a replenishment order is issued to the supplier. A major inventory decision that must be made is how much to order when it becomes necessary to replenish the inventory. The order quantity determines the average inventory of electronic switches that will be on hand during the year.

Before proceeding further, we should reflect a bit on this inventory situation. Even when it is known that 12,000 electronic switches will be needed during the year, the problem of how many to order and stock at a time is not a trivial one. There are a number of possibilities. Why not order an entire year's supply at once? Or would it be better to order 6000 at a time, placing two orders during the year? What about keeping the investment in inventory to a minimum by ordering weekly or even daily? Perhaps the best question would be: How can the alternatives be evaluated to arrive at the best inventory policy? The answer, of course, must come from an analysis of the costs affected by the order-quantity decision.

As discussed earlier, the relevant costs to be considered for the decision are the ordering cost and the carrying cost. Let us assume that a fixed cost of $30 is incurred each time an order is placed with the supplier and that the cost to carry an electronic switch in inventory for a year is estimated to be $2. The following mathematical notation will be useful in summarizing the data and developing the inventory model:

$$\text{Annual demand for the item } D = 12{,}000 \text{ units}$$
$$\text{Purchase price of the item } p = \$10 \text{ per unit}$$
$$\text{Ordering cost } C_O = \$30 \text{ per order}$$
$$\text{Carrying cost } C_K = \$\,2 \text{ per unit per year}$$

Since any order quantity selected will affect both annual ordering cost and annual carrying cost for the item, the best order quantity is the one that minimizes the total of the two cost categories. Expressed on an annual basis, the cost associated with any order quantity for the electronic switch is

$$\begin{array}{c}\text{Average annual} \\ \text{inventory cost (total cost)}\end{array} = \begin{array}{c}\text{average annual} \\ \text{ordering cost}\end{array} + \begin{array}{c}\text{average annual} \\ \text{carrying cost}\end{array}$$

Notice that the annual cost to purchase the item is not considered part of the inventory cost in this situation. Regardless of the size of each replenishment order, 12,000 switches will be purchased over the course of the year at a cost of $10 each. Thus, there will be $120,000 in annual purchase cost, regardless of the inventory policy established. The decision here is concerned solely with how many of the 12,000 electronic switches to purchase with each replenishment order. The annual purchase cost is affected by order sizes only in situations where price discounts are offered for large purchase orders for the item. That is a common situation which we examine after developing and illustrating the basic model.

The mathematical model is developed by translating the total cost expression into mathematical notation. This is nothing more than an algebraic formulation of the problem. For the average annual ordering cost, the expression is

$$\frac{D}{Q}C_O$$

where Q is the order quantity.

The expression D/Q gives the average number of orders that will have to be placed annually if D units of the item are needed during the year and Q units are ordered each time the inventory is replenished. The number of orders placed during the year times the cost per order C_O yields the average annual ordering cost for the item. As discussed earlier, the annual ordering cost decreases as Q gets larger and larger.

The mathematical expression for the average annual carrying cost is

$$\frac{Q}{2}C_K$$

The expression $Q/2$ yields the average inventory of the item that will be on hand over the course of the year if Q units are ordered each time the inventory is replenished and the units are used at a constant rate throughout the year (as we have assumed in the continuous production process). Of course, this is an average figure, so the inventory level of the switches would be greater than $Q/2$ half the time and less than $Q/2$ half the time. Multiplying the average inventory by the cost to carry a unit in stock for a year C_K yields the average annual carrying cost for the item. The expression verifies something we have already established: the cost to carry an inventory of the item increases as the order quantity Q increases.

The general mathematical model for the total cost TC of the order-quantity decision becomes

$$\text{TC} = \frac{D}{Q}C_0 + \frac{Q}{2}C_K \qquad (10\text{-}1)$$

Since the order quantity Q is the only unknown, the expression for our example is

$$\text{TC} = \frac{12,000}{Q}(30) + \frac{Q}{2}(2)$$

Now it is possible to use the model to evaluate any possible order quantity in terms of the average annual inventory cost it would create. This is done by simply substituting the value for Q into the model and computing the total cost. For example, it might be convenient for administrative purposes to place orders with the supplier on a quarterly basis. In that case, each order would be for $12,000/4 = 3000$ switches and the annual inventory cost would be

$$\text{TC} = \frac{12,000}{3000}(30) + \frac{3000}{2}(2)$$
$$= 120 + 3000$$
$$= \$3120$$

TABLE 10-1   Arbitrary ordering strategies

Ordering strategy	Order quantity Q	Average annual ordering cost	+	Average annual carrying cost	=	Average annual inventory cost
Daily	40	$ 9,000		$ 40		$ 9,040
Every other day	80	4,500		80		4,580
Monthly	1,000	360		1,000		1,360
Quarterly	3,000	120		3,000		3,120
Semiannually	6,000	60		6,000		6,060
Annually	12,000	30		12,000		12,030

The decision to order quarterly would be an arbitrary one, and the $3120 represents the annual cost created solely by the policy established to acquire the 12,000 switches that we know will be needed over the year. Some other arbitrary solutions and their corresponding annual costs are tabulated in Table 10-1.

Three observations can be made from the table: (1) The order-quantity decision for this $10 item can generate significant annual cost. If a policy were established to order an entire year's supply at once, the annual inventory-related cost would exceed $12,000! (2) There are large variations among the annual costs for the six policies. Ordering annually (Q = 12,000) is almost 9 times more costly than ordering monthly (Q = 1000). This suggests that a completely arbitrary decision might be much more costly than one based on a cost analysis or comparison. (3) There is a definite pattern to the behavior of annual cost as the order quantity increases. The average annual cost falls sharply and then begins to climb again as Q increases. Figure 10-2 illustrates the relationship between order quantity and annual inventory cost and provides a basis for finding the optimal or least-cost order quantity.

Our six arbitrary solutions are easily identified on the cost curve in Figure 10-2. The general shape of the cost curve is typical of any inventory item. Annual inventory cost is high when very small quantities are ordered because the cost associated with the ordering process must be incurred many times throughout the year. Conversely, annual inventory cost is also high when very large quantities are ordered because of the cost of carrying such a large inventory. There will always be some middle ground where the total cost associated with these two opposing forces is minimized. In Figure 10-2 the most economical order quantity Q_o is clearly the one corresponding to the minimum point on the average annual-cost curve. We could proceed from here to find Q_o by trial and error since we have an expression for annual cost and know from Figure

10-2 that Q_0 is somewhere between our arbitrary solutions of 80 and 1000. However, the trial-and-error approach is not necessary since a general formula exists for finding the economic order quantity.

The EOQ Formula

The general formula for identifying the order quantity that minimizes average annual inventory cost is called the *economic-order-quantity* (EOQ) *formula.* It is derived from the general mathematical model (10-1) by using techniques of differential calculus. The EOQ formula is

$$Q_0 = \sqrt{\frac{2DC_0}{C_K}} \tag{10-2}$$

Notice that use of the formula requires estimates of three parameters, annual demand for the item D, ordering cost C_0, and carrying cost C_K. Also note that the formula exhibits the relationships between (1) ordering cost and Q_0 and (2) carrying cost and Q_0 that we have intuitively suspected. For example, as the cost to place orders for the item C_0 gets larger and larger relative to the other variables, the numerator in the fraction under the radical increases and the value for Q_0 increases. This is consistent with our reasoning that large order quantities are appropriate in sit-

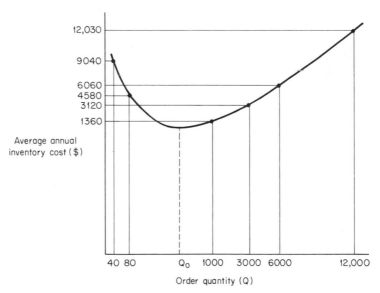

Figure 10-2 Inventory cost curve.

uations where the ordering cost is very high. On the other hand, when the cost to carry the item in inventory C_K is very high, the denominator in the fraction becomes very large and the resulting Q_0 will be small. Again, the mathematics of the formula are consistent with our reasoning that items that are expensive to carry should be ordered in small quantities. In effect, the EOQ formula explicitly trades off the two opposing costs and identifies Q_0. For the example problem, the economic order quantity is

$$\begin{aligned} Q_0 &= \sqrt{\frac{2(12,000)(30)}{2}} \\ &= \sqrt{360,000} \\ &= 600 \end{aligned}$$

When 600 electronic switches are ordered with each replenishment order, average annual inventory cost is

$$\begin{aligned} TC &= \frac{12,000}{600}(30) + \frac{600}{2}(2) \\ &= 20(30) + 300(2) \\ &= 600 + 600 \\ &= \$1200 \end{aligned}$$

The optimal policy would require, on average, that 20 orders (12,000/600) be placed during the year, each order making up a 15-day supply (600/40). Over the course of the year, an average of 300 electronic switches would be in inventory (600/2). Figure 10-3 illustrates graphically how the inventory level for the electronic switch would change over time.

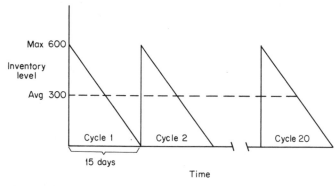

Figure 10-3 The inventory cycle.

There would be 20 inventory cycles (each represented by a triangle) during the year.

Use of the EOQ Formula

The EOQ formula is appropriate only under circumstances that are consistent with our example problem. Therefore, we should briefly point out again the major assumptions made in developing the model. One important assumption implied by the use of formula (10-2) is that the inventory item is used at a constant rate throughout the planning period. This is usually a reasonable assumption in a continuous production system, as we assumed in the example problem. For items whose use is seasonal, it is sometimes possible to select the planning period in such a way that demand for the item is fairly uniform.

Another important assumption made in developing the EOQ formula is that the entire order quantity is received and entered into inventory at the same time. In the example, then, our assumption is that all 600 switches will come in the same shipment and go into inventory together. Also, use of the basic formula implies that the price of the item does not depend on the quantity ordered. This assumption will be relaxed in the next section.

Finally, the model assumes that the inventory manager knows or can accurately estimate demand for the item D, its ordering cost C_O, and its carrying cost C_K. While in reality this may be a difficult task, without estimates of these important variables there is no basis at all for developing a rational inventory policy. The alternative is to establish the order quantity for the item on a completely arbitrary basis, and Table 10-1 illustrates how costly that can be.

The assumptions made in developing the basic EOQ formula make it applicable to probably only a small number of actual situations. However, it does seem to provide ball park solutions in many situations that do not strictly fit the assumptions. Also, it illustrates rather well how an important inventory problem can be analyzed in a quantitative framework. That, of course, is our major purpose here.

It must also be noted that the limitations and assumptions of the EOQ model refer specifically to that particular model, not to the use of inventory models in general. As mentioned earlier, an inventory model can be developed to describe almost any set of circumstances accurately. Of course, the more complicated the circumstances the more complex the mathematics necessary to describe and solve the model. Even so, many of the concepts involved in developing the basic EOQ model will also apply in more advanced models. We now illustrate how the basic model can be applied in situations involving quantity discounts.

THE EOQ MODEL WITH QUANTITY DISCOUNTS

Suppose that the supplier of the electronic switch in the example problem offers to discount the price if a large quantity is ordered. Specifically, let us assume that the following price information is given by the supplier:

Order quantity	Price
Less than 2000	$10.00
2000 or more	9.50

Except for the quantity discount, the situation remains the same as previously described. The question now becomes: How does the discount schedule affect the order quantity decision? The question can be answered by revising the inventory model to account for the new situation. The following variables are relevant to the revised model:

$$\text{Annual demand for the item } D = 12{,}000 \text{ units}$$
$$\text{Purchase price of item } p = \begin{cases} \$10.00 \text{ if } Q < 2000 \\ \$9.50 \text{ if } Q \geqslant 2000 \end{cases}$$
$$\text{Ordering cost } C_O = \$30 \text{ per order}$$
$$\text{Carrying cost } C_K = \$2 \text{ per unit per year}$$

We still want to select the order quantity that minimizes average annual inventory cost. With the quantity discount, however, annual purchase cost as well as ordering and carrying costs for the electronic switch are affected by the order-quantity decision. The appropriate expression for annual inventory cost now becomes

$$\begin{array}{c}\text{Average annual} \\ \text{inventory cost} \\ \text{(total cost)}\end{array} = \begin{array}{c}\text{average annual} \\ \text{ordering cost}\end{array} + \begin{array}{c}\text{average annual} \\ \text{carrying cost}\end{array} + \begin{array}{c}\text{average annual} \\ \text{purchase cost}\end{array}$$

To develop the inventory model, the cost components must again be expressed in mathematical notation. This has already been done for the ordering and carrying costs. For the average annual purchase cost the mathematical expression is simply pD (where p depends on the order quantity).

The general inventory model for the quantity discount situation can then be expressed as

$$TC = \frac{D}{Q} C_O + \frac{Q}{2} C_K + pD \tag{10-3}$$

and for the example problem

$$TC = \frac{12,000}{Q} (30) + \frac{Q}{2} (2) + p(12,000)$$

At this point, the model could be used to evaluate any order quantity by substituting the appropriate values for Q and p (p will be either $10 or $9.50 depending on Q) and computing average annual inventory cost. Since the model consists of a single equation with two unknowns, it is not possible to derive a formula that yields Q_0 directly, as was done previously. However, it is not necessary to proceed by trial and error either. We have already determined that in the absence of any quantity discount the best order quantity is 600. This information is provided from the EOQ formula (10-2). When Q equals 600, the price is $10 and the average annual cost for ordering, carrying, and purchasing [from formula (10-3)] is

$$TC = \frac{12,000}{600} (30) + \frac{600}{2} (2) + 10(12,000)$$
$$= 600 + 600 + 120,000$$
$$= \$121,200$$

To determine whether the order quantity should be increased to take advantage of the discount, only one cost comparison need be made. It is simply a matter of comparing the $121,200 figure with the average annual cost when an order quantity of 2000 is used to obtain the discounted price. When Q equals 2000, the price is $9.50 and annual cost is

$$TC = \frac{12,000}{2000} (30) + \frac{2000}{2} (2) + 9.50(12,000)$$
$$= 180 + 2000 + 114,000$$
$$= \$116,180$$

In effect, the savings in annual purchase cost ($120,000 − $114,000) more than offset the cost associated with carrying the additional stock, and the order quantity of 2000 is more economical.

Another Price Option

It is not uncommon for a supplier to offer several price breaks at successively larger quantities. For example, the supplier of the electronic switch in our illustration might reduce the price to $9.40 if 6000 units or more are ordered. Would it be economically advantageous to increase the order quantity even further to get the $9.40 price? It is simply a matter of using the model to compute annual cost when Q equals 6000 and p equals $9.40:

$$TC = \frac{12,000}{6000}(30) + \frac{6000}{2}(2) + 9.40(12,000)$$
$$= 60 + 6000 + 112,800$$
$$= \$118,860$$

In this case, the carrying cost on the additional inventory that would be created by ordering 6000 switches more than offsets the savings on the purchase price, and the best inventory policy is to remain with the 2000 order quantity. Order quantities in between 600, 2000, and 6000 need not be evaluated since they would merely increase carrying cost without generating additional savings in purchase cost. Only at the price breaks is the purchase cost reduced. Figure 10-4 illustrates the inventory cost curve for this quantity discount problem. Notice that the price discounts cause the curve to be "kinked" at order quantities of 2000 and 6000. As we have calculated, the minimum point on the curve occurs when the order quantity is 2000.

To summarize the procedure, the EOQ formula (10-2) is used to develop a starting point by identifying the optimal order quantity when

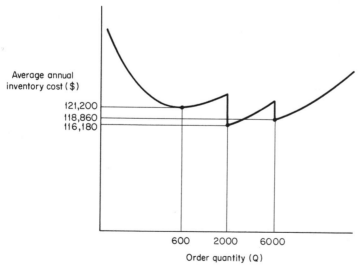

Figure 10-4   Inventory cost curve with quantity discounts.

the price discounts are ignored. Then, the revised inventory model (10-3) is used to evaluate the larger order quantities corresponding to the price breaks. The best order quantity is the one that minimizes annual inventory cost.

A SAFETY-STOCK/REORDER-POINT MODEL

In Figure 10-3 the behavior of the inventory level for the electronic switch is illustrated when each replenishment order is for 600 units. In the absence of any quantity discounts, 600 was computed as the optimal order quantity. In this section we refer back to the original example and assume that no quantity discounts are offered. Each triangle in Figure 10-3 represents an inventory cycle, and the downward sloping line shows the depletion of the inventory over time. At some point as the stock is reduced, a decision must be made to issue a replenishment order for an additional 600 switches. Expressed in terms of the number of units remaining in stock, the point at which an order is placed is called the *reorder point.*

Establishing the appropriate reorder point for the electronic switch in our example would be a simple matter if delivery time (often called *lead time*) from the supplier were known with certainty. For example, if the supplier *always* gave 3-day delivery service, replenishment orders would simply be placed when a 3-day supply of switches was left in stock. Since the example problem deals with an item used in continuous production at the rate of 40 per day, the reorder point would be 120. In many situations both use of the item and lead time from the supplier are subject to uncertainties, and the selection of the appropriate reorder point can be very difficult.

Selecting the reorder point is important because two types of costs are affected by the decision. If the reorder point is too low, not enough stock will be left to last through the delivery period and the result will be a stock-out. To avoid the cost associated with a stock-out, the reorder point can be established so that the remaining stock is more than enough to last through the expected delivery time. The extra stock, called *safety stock,* provides protection against unusually long lead times (or a higher rate of use than expected). However, the safety stock will normally not be used and therefore must be carried in inventory throughout the year. The opposing forces created by the stock-out cost and the extra carrying cost can be formulated in a mathematical model to identify the most economical reorder point.

To illustrate the formulation of the model, let us assume that the sup-

TABLE 10-2   *Probability distribution for use during delivery period*

Delivery period, days	Use during delivery period (40 per day)	Probability
1	40	.10
2	80	.20
3	120	.47
4	160	.10
5	200	.09
6	240	.04
		1.00

plier of the electronic switch in the example usually makes delivery in 3 days but the delivery time has been known to vary anywhere from 1 to 6 days. Suppose that historical data concerning the delivery service of the supplier are examined and the probability distribution in Table 10-2 is constructed.

The table implies that lead time can never be less than 1 day or longer than 6 days. Since it is known that 40 switches will be needed each day, the probability distribution describes how use of the electronic switches may vary while an order is outstanding. The expected value of the distribution is 120, computed as follows:

$$\text{Expected use during lead time} = 40(.10) + 80(.20) + 120(.47)$$
$$+ 160(.10) + 200(.09) + 240(.04)$$
$$= 4 + 16 + 56.4 + 16 + 18 + 9.6$$
$$= 120$$

If the reorder point is established at 120, orders are placed when the remaining stock is just enough to meet expected use during lead time. This is equivalent to carrying no extra, or safety, stock for protection against the possibility that 160, 200, or even 240 switches will be needed before the order arrives. If orders are placed when 160 switches remain in stock, the probability of running out decreases but additional carrying cost is incurred for the safety stock of 40 units.

Now assume that a stock-out results in an emergency order to the supplier that costs $60 (recall that the regular ordering cost for the electronic switch is $30). The emergency order must be placed regardless of the exact number of units short, so the $60 cost refers to the occurrence of a stock-out. The annual carrying cost for the item is $2, and we have determined from the earlier EOQ analysis that 600 switches should be ordered

at each of 20 replenishment cycles during the year. The following notation will be used in the reorder-point-safety-stock model:

$$\text{Stock-out cost } C_S = \$60 \text{ per occurrence}$$
$$\text{carrying cost } C_K = \$2 \text{ per unit per year}$$
$$\text{Probability of a stock-out } P = \text{Determined from Table 10-2}$$
$$\text{for a given reorder point}$$
$$\text{Number of orders placed per year } N = 20$$

The best reorder point is the one that minimizes the total of the stock-
out and carrying costs it creates. Expressed on an annual basis, the total
cost affected by the reorder-point decision is

$$\begin{matrix} \text{Average annual cost} \\ \text{(total cost)} \end{matrix} = \begin{matrix} \text{average annual} \\ \text{stockout cost} \end{matrix} + \begin{matrix} \text{average annual} \\ \text{carrying cost} \\ \text{(for the safety stock)} \end{matrix}$$

In mathematical notation, the average annual stockout cost is

$$C_S NP$$

where P is determined by the reorder point. Since P represents the probability of a stock-out each time an order is placed, NP yields the expected
number of stock-outs encountered during the year and $C_S NP$ expresses
the cost in dollars.

The mathematical expression for average annual carrying cost for the
safety stock is simply

$$C_K S$$

where S is the safety stock.

On the average, the units maintained in safety stock will not be needed
(they are there to protect against variations from the average), so each of
the S units will generate C_K in annual carrying costs.

The general mathematical model then becomes

$$TC = C_S NP + C_K S \tag{10-4}$$

and for the example problem,

$$TC = 60(20)P + 2S$$

The model can now be used to evaluate any possible safety stock (reorder point) in terms of its impact on annual cost. Note that the model

consists of one equation with two unknowns, and so it is not possible to develop a single formula that yields the optimal reorder point directly. However, only four possibilities need to be evaluated in this case. We would never consider letting the stock fall below 120 switches before placing an order, so the possible reorder points are 120, 160, 200, and 240.

To illustrate the use of the model, consider a reorder point of 120, which is exactly the expected use of the item during lead time. This means that no safety stock is carried at all ($S = 0$). From Table 10-2, a reorder point of 120 will result in a stock-out if 160 or 200 or 240 switches are needed before the order arrives. The sum of those probabilities (.10 + .09 + .04 = .23) represents the probability of a stock-out, P, at each replenishment order. Thus, if the reorder point is 120, $S = 0$, and $P = .23$, and

$$TC = 60(20)(.23) + 2(0)$$
$$= 276 + 0$$
$$= \$276$$

Computations for the other three possible reorder points are summarized in Table 10-3.

The table shows that average annual cost is minimized if orders are placed when the inventory level falls to 200 switches. The safety stock of 80 units reduces the probability of a stock-out to .04 and provides a balance between the opposing costs. With this policy, it would be expected that 96 of every 100 orders would arrive before the remaining stock is completely depleted. The table also illustrates a characteristic of most inventory items. It is usually not economically feasible to attempt to eliminate the possibility of a stock-out completely. The safety stock required to provide complete protection against stock-outs (120 units in this case) usually generates prohibitively high carrying cost.

TABLE 10-3   Reorder-point levels and average annual cost

Reorder point	Safety stock S	Probability of a stock-out P	Average annual stock-out cost	+	Average annual carrying cost	=	Average annual cost
120	0	.23	$276		$ 0		$276
160	40	.13	156		80		236
200	80	.04	48		160		208
240	120	0	0		240		240

CONCLUSION

Inventory models provide a quantitative basis for making inventory decisions. Many such decisions involve costs that influence the decision maker in opposite directions, and the inventory model serves as a framework in which explicit cost trade-offs can be made. The development of an inventory model often provides insights into the problem that are far more valuable than merely getting the best solution. Although the basic models presented in this chapter are elementary, they serve to illustrate the potential uses and benefits of inventory models.

CHAPTER 11

Waiting-Line Models

Waiting lines are common occurrences in everyday life. All of us experience waiting lines at checkout counters in grocery stores, traffic lights at highway intersections, drive-in windows at banks, and numerous other situations. In fact, it is virtually impossible to go through an entire day without encountering some type of system that creates or generates a waiting line. The study of waiting-line systems, or *queuing systems,* has been of interest to operations researchers since around 1910, when A. K. Erlang studied automatic telephone exchange systems and the resulting delay, or waiting time, experienced by callers. Today, a fairly substantial body of knowledge exists to help managers estimate the impact that operating policies and procedures will have on waiting-line systems.

A waiting-line system is defined as any system to which arrivals come for a particular service and must wait in line if the facility that provides the service is busy. The most obvious examples (which we can all identify with) involve people as arrivals waiting for service in stores, gas stations, restaurants, banks, and so forth. However, there are many industrial examples of waiting-line systems that, while less obvious, may have an important impact on overall operating efficiency. For example, trucks arriving at a loading dock must sit idle and wait if the dock is already full. In a computer system, submitted jobs form a waiting line to be processed by the computer according to some predetermined priority rules. As another example, secretaries who use a centrally located duplicating machine must stand idle and wait if the machine is already in use when they arrive.

In all these situations there exists a trade-off in costs. First, there is a cost associated with arrivals having to wait in line. For idle secretaries and truck drivers, the waiting cost is simply the wages they are paid while being idle. For customers in a grocery store or bank, the waiting cost is difficult to measure, but excessively long waiting lines can lead eventually to lost sales and damaged goodwill.

Second, there is a cost associated with providing service. In any system, waiting lines can be reduced by simply increasing the service capability, that is, by hiring additional dock workers to load and unload trucks, or by purchasing a second duplicating machine, or by establishing an additional checkout counter. Of course, the additional manpower and/or equipment creates higher operating costs, and the knowledgeable manager will attempt to establish policies that find a happy medium between service costs and waiting costs.

The above discussion implies that it is not usually feasible to try to eliminate waiting lines entirely. This is true since the resulting service costs would be prohibitively high if we tried to ensure that nobody would ever have to wait in line. Waiting-line models are useful because they provide a systematic means of estimating how management decisions affect the nature of the waiting line.

TECHNICAL CHARACTERISTICS
OF WAITING-LINE SYSTEMS

A waiting line or queue develops because of the relationship and interaction between two processes, *arrivals* and *service*. In many cases, arrivals occur on a random basis. For example, the number of customers coming into a store in a given period is a random variable that may be completely outside the control of the store owner. In similar fashion, the arrival of trucks at a loading dock or secretaries at a duplicating facility may occur without any apparent pattern. In other cases, it is possible to influence the arrival process directly. For example, the promotion of special sales during selected times or the establishment of a "happy hour" are both examples of attempts to alter the basic nature of the arrival process. In the extreme, the arrival process might be made completely deterministic by requiring appointments for all arrivals.

Service is also usually viewed as a random process in the sense that no two arrivals require the same service time. The time required to provide service at a grocery checkout counter depends on the number of items purchased by the customer. In a bank drive-in window, the service time varies with the complexity of the transaction. Similarly, loading or unloading time at a shipping dock depends on the size of the shipment and the nature of the items. In some systems, efforts can be made to make the service time more uniform. For example, an express lane at a grocery checkout counter is designed to process arrivals with essentially the same needs, for example, 10 purchased items or less, and level out service times. Sometimes during the rush hours banks designate certain windows for check cashing only. Again, the purpose is to reduce the variation in service times.

Both the arrival process and the service process in any waiting-line system can be described in the form of a probability distribution. For example, if we were to observe the number of arrivals per hour in a system over a long period of time, the result would be a distribution of numbers with a specific mean and standard deviation. In analyzing the system, an operations researcher would attempt to identify the nature or shape of the distribution and estimate its mean and standard deviation. Similarly, the service time required by each arrival also forms a probability distribution with its own mean and standard deviation. It is the interaction between these two probability distributions that causes waiting lines. By identifying the nature of the distributions it is possible to make meaningful predictions about the average length of the waiting line or the average time each arrival will have to wait.

Other Technical Points

Every waiting-line system also must have some type of priority rule governing the order in which waiting units are served. This is called the *queue discipline*. In many waiting-line systems, the queue discipline is simply first come, first served. That is, the earliest arrival (the unit at the head of the waiting line) has first priority when the service facility becomes available. This, of course, is the queue discipline that guides the familiar waiting lines in banks, retail stores, restaurants, and many other systems encountered on a daily basis.

Some waiting-line systems in industry might employ a queue discipline different from the first come, first served rule. For example, in a job shop a waiting line might be formed by orders that are awaiting processing on a particular piece of equipment. Here, the queue discipline might be "earliest due date first." That is, the order that has the earliest promised delivery date has first priority for processing on the piece of equipment in question. Another possible queue discipline would be "shortest processing time first," where the order requiring the shortest time is given first priority.

The queue discipline is an important characteristic of a queuing system since it affects the nature of the waiting line. In building a descriptive model of a waiting-line system, it is extremely important to identify the queue discipline accurately. In some cases, it might be appropriate to construct a model to evaluate alternative queue disciplines in order to identify the one that has the desired effect on the waiting-line system.

Another important technical aspect of any waiting-line system is whether the system is *single-* or *multichannel*. In a single-channel system, one waiting line forms to enter a single service facility. This would be the case in a grocery store that had only one checkout counter or a bank that had a single drive-in window. Other single-channel systems include a shipping and receiving area consisting of a single loading dock or a central duplicating room containing only one copying machine. In all these examples, a single line forms waiting the availability of one service area. All arrivals go through the same waiting line and service facility; it is simply a matter of how long, if at all, each arrival will have to wait.

Multichannel systems are those in which more than one service facility exists and waiting units may enter the first service facility that becomes available. For example, a bank that operates several drive-in windows would constitute a multichannel system since the car at the head of the waiting line could enter the first vacant service window. A shipping and receiving area made up of several loading docks would also be a multichannel system because a waiting truck could pull up to the first vacant dock. Similarly, a central duplicating facility with two identical copying machines would constitute a multichannel system. In all these cases, each arrival goes through the waiting line and one of the service facilities. Which facility serves each arrival depends on the chance interactions between the arrival and service processes.

The arrival process, service process, queue discipline, and channel structure represent the technical characteristics of a waiting-line system. The first step in developing a waiting-line model is to identify these characteristics for the system under study. The mathematical sophistication necessary to develop a waiting-line model depends greatly on the nature of the technical aspects discussed in this section. As we shall demonstrate, certain types of waiting-line systems can be described and analyzed with models that require only elementary mathematics.

PERFORMANCE CHARACTERISTICS OF WAITING-LINE SYSTEMS

Before illustrating the development and application of a waiting-line model, let us define some performance measures frequently used to evaluate queuing systems. The performance characteristics really describe how the system operates and are often used as a basis for evaluating alternative policies or operating procedures. In effect, the purpose of a waiting-line model is to provide some insights into the expected performance of the system. The following are important performance characteristics for any waiting-line system:

Utilization The percentage of time the service facility is expected to be busy. This is a measure of the productive time of the teller in the bank drive-in window, the clerk at the grocery checkout counter, the workers at the shipping and receiving dock, and so forth. It also provides a measure of the probability that an arrival will have to wait for service. When utilization is very low, the server is idle a high percentage of the time and chances are good that an arrival will not have to wait at all. Conversely, high utilization means a busy service facility and a low probability that a new arrival can be served immediately. The utilization measure, of course, varies between 0 and 100 percent.

Average number waiting The average number of units in the queue or waiting line. This is an estimate of the average size of the waiting line.

Average number in the system The average number of units waiting plus the average number of units being served. This is an estimate of the number of units in the system at any given time.

Average waiting time The average time that an arrival has to wait before being served. In any system, some arrivals will be served right away while others will experience a waiting line.

Average time in the system The average waiting time plus the average service time for each arrival into the system. This is an estimate of the average time required to go through the system.

AN ELEMENTARY WAITING-LINE MODEL

Let us suppose that a manufacturing facility contains a tool issue room, where machinists obtain the tools necessary for their work. Since the tools are precision instruments that are very expensive, an accurate accounting of their use is required. A single attendant operates the toolroom, issuing the requested tools and maintaining a use log. A machinist arriving at the toolroom must wait in line if the attendant is busy serving another machinist. This situation is one of the classic examples of a waiting-line system in industry.

In order to develop a descriptive model of this waiting-line system, let us first identify its technical characteristics. First, how can we describe the process by which machinists arrive at the toolroom? To answer that question, suppose a study was conducted to observe the actual arrival rate of the machinists. The number of arrivals per hour is recorded over a period of time, and the resulting data are summarized in Figure 11-1. In some hours no machinists at all came to the toolroom, whereas in other hours

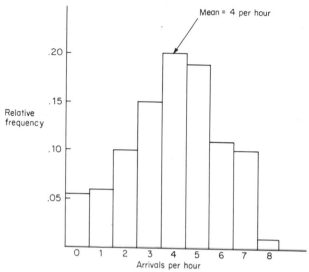

Figure 11-1   Arrivals per hour at the toolroom.

as many as eight machinists needed to check out a tool. On the average, four machinists arrived at the toolroom each hour.

The axis labeled "Relative frequency" in Figure 11-1 indicates the proportion of the total observations that each number of arrivals per hour was observed. If the observations were made over a period considered typical of normal operations, the relative frequencies may be viewed as probabilities and the distribution treated as a probability distribution. Furthermore, the nature of the distribution in Figure 11-1 conforms to that of the Poisson distribution. Thus, in describing the arrival process, it can be said that the number of machinists arriving at the toolroom each hour is Poisson-distributed with a mean of 4.

The Poisson distribution has been found to describe the arrival process in many waiting-line systems accurately. Since the Poisson is an easy distribution to work with and seems to apply to a variety of queuing situations, it is not surprising that many mathematical waiting-line models are based on the assumption of a Poisson arrival process.

Having identified and described how machinists arrive at the toolroom, let us now focus our attention on the service process. By observing the toolroom over some typical period and recording the actual service times, the nature of the service distribution can also be identified. Suppose the observed service data are as summarized in Figure 11-2. The data show that the observed times required to check out a tool varied from 1 to 15 minutes, the average time being 5 minutes.

The histogram in Figure 11-2 shows the frequency with which each

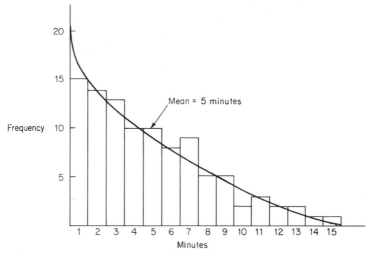

Figure 11-2   Service times at the toolroom.

service-time interval was recorded over the observation period. Since time is a continuous variable, the service distribution is represented by the continuous line drawn in the figure. A continuous distribution of this general form is referred to as an *exponential* distribution. In describing the nature of the service process, then, it can be said that the service time is exponentially distributed with a mean of 5 minutes.

Empirical research has shown that the exponential distribution accurately describes the service process in many queuing systems. The statistical properties of the exponential distribution are not really important to us here, except to note the general "curved-line" shape that characterizes the distribution. The models we examine in this chapter assume that service times are adequately represented by an exponential distribution.

With this knowledge of the arrival and service processes, we need only identify the queue discipline and channel structure to describe the technical characteristics of the system. Let us assume that priority for service is based on time of arrival, so that the queue discipline is first come, first served. Since there is only one toolroom with a single waiting line, the system is classified as a single-channel system. Before proceeding let us summarize the technical description of this waiting-line system:

1. Poisson-distributed arrival process (with a mean of 4 per hour)

2. Exponentially distributed service times (with a mean of 5 minutes)

3. First come, first served queue discipline

4. Single-channel system

BASIC QUEUING FORMULAS

Operations researchers have developed a mathematical model for waiting-line systems with the characteristics described in the toolroom example. The model consists of a set of formulas that estimate the important performance characteristics of the system.

To use the standard formulas, one must establish values for the following variables:

$$\lambda = \text{average arrival rate per time period}$$
$$\mu = \text{average service rate per time period}$$

In the example problem, these estimates were assumed to be empirically determined by direct observation of the process. Thus,

$$\lambda = 4 \text{ per hour}$$
$$\mu = \frac{60 \text{ minutes}}{5 \text{ minutes}} = 12 \text{ per hour}$$

On the average, then, 4 machinists arrive each hour at the toolroom, which is capable of serving 12 machinists each hour. Note that λ and μ represent the *rate* of arrival and *rate* of service, respectively, over some common time period. Since the service capability greatly exceeds the arrival rate, we might intuitively suspect that very few machinists will experience a waiting line at the toolroom. The following standard queuing formulas describe the nature of the system:

$$\text{Utilization } U = \frac{\lambda}{\mu} \tag{11-1}$$

$$\text{Average number in waiting line } W = \frac{\lambda^2}{\mu(\mu - \lambda)} \tag{11-2}$$

$$\text{Average number in system } S = U + W \tag{11-3}$$

$$\text{Average waiting time } T_W = \frac{\lambda}{\mu(\mu - \lambda)} \tag{11-4}$$

$$\text{Average time in system } T_S = T_W + \text{average service time} \tag{11-5}$$

For the toolroom

$$U = \frac{\lambda}{\mu} = \frac{4}{12} = .33$$

$$W = \frac{\lambda^2}{\mu(\mu - \lambda)} = \frac{4^2}{12(12 - 4)} = .17$$

$$S = U + W = .33 + .17 = .50$$

$$T_W = \frac{\lambda}{\mu(\mu - \lambda)} = \frac{4}{12(12 - 4)} = .0417 \text{ hour} = 2.5 \text{ minutes}$$

$$T_S = T_W + \text{average service time} = 2.5 \text{ minutes} + 5 \text{ minutes}$$
$$= 7.5 \text{ minutes}$$

Thus, one would expect the attendant in the toolroom to be busy only one-third of the time ($U = .33$). Stated another way, there is a .67 probability that a machinist arriving at the toolroom will not experience a waiting line. The average number of machinists waiting in line W is .17, and the average number of machinists in the system S is .50. Remember that these figures represent long-run averages. If we were to make a large number of observations of the toolroom, we would find that many times there is no one waiting, while on other occasions there may be two, three, or even more machinists in line. The average number waiting would be .17.

The calculations also reveal that the average waiting time T_W is 2.5 minutes and the average time it takes a machinist to go through the waiting line and check out a needed tool T_S is 7.5 minutes. These data not only provide important insights into the nature of the waiting-line system but also provide a basis for evaluating management decisions that might affect its operation.

A Cost Analysis

Suppose the attendant in the toolroom is paid $1.75 per hour and machinists in the shop earn an average of $7.50 per hour. A proposal has been made to hire a helper to assist the toolroom attendant. Management must determine the impact that the helper would have on the performance of the system and the costs of operation.

At first glance it would seem inappropriate to hire a second worker for the toolroom. After all, the current attendant is now idle two-thirds of the working day. How could it possibly be economically advantageous to provide him with an assistant? The answer, of course, lies in the fact that the system must be evaluated on the basis of both service *and* waiting costs. Undoubtedly, the addition of a helper in the toolroom will increase service costs. The increased ability to serve the machinists, however, should result in faster service times and fewer highly paid machinists standing in line. Both factors must be considered to evaluate the proposal. Furthermore, the queuing formulas provide a means of estimating the performance characteristics of the system if the helper is hired.

Suppose that management estimates that the addition of a helper would decrease the average service time at the toolroom to 3 minutes, still exponentially distributed. The proposed system would have the attendant and helper working together to serve each arriving machinist. While one logs the transaction, the other would go back into the storeroom to get the needed tool. In this fashion, the system would still operate as a single-channel facility and the basic queuing formulas still apply.

TABLE 11-1   *Performance characteristics of the toolroom*

	Single attendant $\lambda = 4$ $\mu = 12$	Attendant and helper $\lambda = 4$ $\mu = 20$
Utilization $U = \dfrac{\lambda}{\mu}$	.33	.20
Average no. waiting $W = \dfrac{\lambda^2}{\mu(\mu - \lambda)}$	.17	.05
Average no. in system $S = U + W$	.50	.25
Average waiting time $T_W = \dfrac{\lambda}{\mu(\mu - \lambda)}$	.0417 hour (2.5 minutes)	.0125 hour (.75 minute)
Average time in system $T_S = T_W + $ average service time	.1250 hour (7.5 minutes)	.0625 hour (3.75 minutes)

With the addition of a helper the parameters of the waiting-line system would become

$$\lambda = \text{arrival rate} = 4 \text{ per hour}$$

$$\mu = \text{service rate} = \frac{60 \text{ minutes}}{3 \text{ minutes}} = 20 \text{ per hour}$$

The performance characteristics of the proposed system are computed and summarized in Table 11-1. Also shown are the previously computed characteristics of the present operation.

As we suspected, the faster service provided by the attendant and helper reduces the average number of machinists waiting and the average time required to go through the system. Yet, as noted earlier, these reductions are gained at the expense of higher service costs. To determine the desirability of employing the helper, we must compare the overall costs (service costs plus waiting costs) of the current system with the proposed system. This is done in Table 11-2.

Table 11-2 shows that the hourly service cost generated by the toolroom would double if the helper is hired (assuming the helper is also paid $1.75 per hour). However, the average hourly cost of the nonproductive machinists would decline from $3.75 to $1.88 as a result of the faster service. The calculation of hourly waiting cost is based on the average number of machinists in the queuing system (and therefore not engaged in productive work) at any given hour. The values of .50 and .25 are taken from the calculations in Table 11-1. Overall, hourly costs of operating the toolroom decrease by $.12 if the helper is hired.

TABLE 11-2   *Hourly cost comparisons*

	Hourly service cost	+	Hourly waiting cost	=	Total hourly cost
Single attendant	1(1.75) = $1.75		.50(7.50) = $3.75		$5.50
Attendant and helper	2(1.75) = 3.50		.25(7.50) = 1.88		5.38

Clearly, here is a case where inexpensive labor is being traded off against highly specialized, expensive labor. It is simply more economical to have idle toolroom attendants than it is to have idle craftsmen. It would have been a serious conceptual error to dismiss the proposal to hire a helper for the toolroom simply because the existing attendant is already idle most of the time. In fact, when the new helper is added, the utilization factor computed in Table 11-1 shows that the toolroom crew would expect to be busy only 20 percent of the time. It might be possible that they can be given other duties to perform when there are no machinists in need of a tool. However, that would not have to be the case; the new helper would still be cost-effective if he and his partner sat in the toolroom and played cards most of the day!

Without some type of waiting-line model, it would be almost impossible to estimate the effect the helper would have on the system. The mathematical model takes the immediate estimate that the average service time at the toolroom would decrease from 5 to 3 minutes and translates it into meaningful projections about the entire waiting-line system. Such data provide extremely important insights for the decision maker. The following example provides a second illustration of the use of waiting-line models in business decision making.

ANOTHER EXAMPLE

Suppose a large insurance company plans to establish a central duplicating room where secretaries from the various offices in the building can do all their copying work. The company plans to install a single, top-quality duplicating machine in the room and make it available to the secretaries as needed on a first come, first served basis. Management estimates that the number of secretaries arriving at the duplicating room will average nine per hour and be adequately described by a Poisson distribution. The secretaries are paid an average of $4 an hour.

Management must select one of three alternative copying machines to install in the duplicating room. The machines vary in speed of operation and monthly cost, as shown in Table 11-3. The average service times are

TABLE 11-3   Monthly machine costs and service capabilities

Copying machine	Monthly cost	Average service time, minutes
1	$ 75	4
2	100	3
3	375	2

estimated for the typical duplicating job, and it is assumed that variation from the average for each machine is adequately represented by an exponential distribution. As Table 11-3 indicates, monthly costs increase as the service capability increases.

If the central duplicating room is viewed as a queuing system, the criterion for selection of the appropriate copying machine should be obvious. The arrival rate of the secretaries has already been estimated (nine per hour) and the service rate will depend on the particular machine selected. The waiting costs in the system are defined as wages paid to secretaries who are away from their desks to perform duplicating work. The service cost is the monthly cost of the copying machine. By installing the fastest machine available, the cost of "idle" secretaries would be minimized. On the other hand, by installing the slowest and cheapest machine, the monthly service cost would be minimized. Clearly, the appropriate objective is to install the machine that minimizes *total systems costs* (waiting costs plus service costs).

TABLE 11-4.   Performance characteristics of three copying machines

	Copying machine		
	1 $\lambda = 9$ $\mu = 15$	2 $\lambda = 9$ $\mu = 20$	3 $\lambda = 9$ $\mu = 30$
Utilization $U = \dfrac{\lambda}{\mu}$	.60	.45	.30
Average no. waiting $W = \dfrac{\lambda^2}{\mu(\mu - \lambda)}$	.90	.37	.13
Average no. in system $S = U + W$	1.50	.82	.43
Average waiting time $T_W = \dfrac{\lambda}{\mu(\mu - \lambda)}$	.10 hour (6 minutes)	.04 hour (2.4 minutes)	.014 hour (.84 minute)
Average time in system $T_S = T_W +$ average service time	.17 hour (10 minutes)	.09 hour (5.4 minutes)	.0473 hour (2.84 minutes)

To develop an estimate of total systems costs, we first estimate the performance characteristics of the system under each of the three copying machines. Formulas (11-1) to (11-5) are applied to yield the results shown in Table 11-4. It should be no surprise that the average number of secretaries in the system decreases, along with the average time required to go through the system, as the service capability of the machine increases. For example, if machine 1 is installed, the average service rate is 15 per hour (60/4) and there would be an average of .9 secretary waiting and 1.5 secretaries in the system at any given time. With machine 1, the average waiting time would be 6 minutes and the average time required for a secretary to complete a duplicating task would be 10 minutes.

For machine 2 the service rate becomes 20 per hour and the average number of secretaries waiting and in the system decreases to .37 and .82, respectively. With machine 2 it would only be expected to take 5.4 minutes to go through the system. Finally, machine 3 increases the service rate to 30 per hour and the relevant performance statistics decrease even further. Now, only .43 secretary would be in the system on the average, and the required time to negotiate the waiting line and complete a duplicating job would decrease to 2.84 minutes.

The cost comparisons can easily be made from the data in Table 11-4. Since the costs of the copying machine are expressed on a monthly basis, let us compare the three alternative systems on the basis of average monthly costs. We will assume that there are 20 working days in a month. As mentioned earlier, the service cost for each system is simply the monthly cost of the duplicating machine. The waiting cost for each system can be determined from the performance characteristics in Table 11-4. If waiting cost refers to the cost of secretaries being away from their desks to perform a duplicating task, we are interested in the average number of secretaries in the system S at any given time. Multiplying the average number in the system by \$4 yields the hourly waiting cost for the system, and the hourly waiting cost times 160 hours (8 hours per day for 20 days) yields the average monthly waiting cost. These computations are summarized in Table 11-5.

As Table 11-5 indicates, monthly costs will be minimized if copying machine 2 is installed in the duplicating room. With this machine, there

TABLE 11-5   Monthly systems costs

Copying machine	Monthly service cost	+	Monthly waiting cost	=	Total monthly cost
1	\$ 75		1.5(4)(160) = \$960		\$1035.00
2	100		.82(4)(160) = 524.80		624.80
3	375		.43(4)(160) = 275.20		650.20

would be an average of .82 secretary in the system at any given time, and the monthly waiting cost would be $524.80. Adding in the $100 monthly machine cost yields a total of $624.80. Machine 1 has lower service costs, but its small service capability would result in excessively high labor costs for the waiting secretaries. Machine 3 would keep the waiting line small, but it generates high service costs. Our analysis shows that machine 2 represents the best balance between the costs of providing the duplicating service and the costs of secretaries waiting to get access to the machine.

CONCLUSION

Waiting-line models provide a systematic means for estimating the important performance characteristics of a system and making management decisions that balance service costs and waiting costs. In this chapter, we have presented and illustrated a basic waiting-line model. It should be emphasized that the mathematical relationships described by formulas (11-1) to (11-5) pertain only to waiting-line systems that are similar to the illustrations in this chapter. Specifically, the formulas are appropriate for waiting-line systems that:

1. Are single channel in nature

2. Operate on a first come, first served basis

3. Have Poisson-distributed arrival rates

4. Have exponentially distributed service times

Many real-world systems meet the above conditions and can be analyzed with the basic queuing model just illustrated. However, there are also many types of systems that consist of multiple channels, more complicated queue disciplines, and/or arrival and service processes that vary from the assumptions here. Analytic models can be developed for many of these more complicated situations as well. While the required mathematics may be more advanced, many of the basic concepts are the same. In extremely complex waiting-line systems, a technique known as simulation (see Chapter 12) is often used to gain insight into the performance characteristics of the system.

A major benefit of a waiting-line model is that it forces the decision maker to take a systems point of view. By illustrating how the arrival process and service process interact to cause waiting lines, a mathematical model allows the decision maker to see how the total system will be affected by alternative policy decisions. In this fashion, many cause-and-effect relationships can be learned from the model and used to establish efficient operating procedures.

CHAPTER 12

Simulation

Simulation is perhaps the most versatile and powerful of the quantitative techniques available to managers. Unlike many of the analytic methods presented in previous chapters, simulation can be used in almost any decision-making situation. This method of analysis has found increasing acceptance among management scientists in recent years.

To simulate is to imitate the essential characteristics of something in order to explain or predict its behavior. In a business situation, the object of a simulation may be the inventory level of a commodity, the time required to complete a project, the service provided by a vendor, the work attendance of employees, or virtually any other phenomenon relevant to a management decision. In any of these situations, a model is developed that identifies the important variables and how they are related in the system under study. Typically, the model is expressed as a mathematical equation (or set of equations) or a flowchart. The simulation analysis consists of a series of trial-and-error experiments on the model to see what happens under a variety of conditions and circumstances. The distinction between simulation techniques and the analytic techniques discussed in earlier chapters is important enough to warrant further discussion.

ANALYTIC MODELS VERSUS SIMULATION MODELS

An analytic model is one that is "solved" directly by the use of standard mathematical or statistical techniques. For example, linear-programming models fall into the analytic category. A linear-programming model is expressed in the form of an objective-function equation and a set of constraint equations or inequalities. By applying a standard set of mathematical procedures (known as the simplex algorithm), the solution that maximizes or minimizes the objective function is systematically identified. A trial-and-error approach is not necessary since any linear-programming model can be optimized with the simplex algorithm. However, as explained in Chapter 8, only certain types of problems lend themselves to linear-programming analysis.

Another familiar example of an analytic model is the economic-order-quantity (EOQ) model. Recall from Chapter 10 that the inventory model is expressed as a mathematical equation relating the quantity of the material ordered to its annual ordering and carrying costs. From the mathematical statement of the model, a general formula [see formulas (10-1) and (10-2)] derived using differential calculus identifies the most economical order quantity. Again, however, certain assumptions are made about the nature of the inventory system, and the analytic model is appropriate only under circumstances that are reasonably consistent with those assumptions.

In similar fashion, the PERT model presented in Chapter 9 and the waiting-line model illustrated in Chapter 11 are analytic in nature. From our previous discussions of all these techniques, two generalizations can be made about analytic models: (1) Analytic models are typically easy to use. Generally, an algorithm or computational formula is available to solve the model. For frequently used models, the general formulas and algorithms are often preprogrammed in computer software packages. In those cases, a manager need only recognize that a decision situation can be accurately described by a particular type of model and provide the necessary data inputs. (2) Analytic models invariably require assumptions that simplify the problem. The simplifications are necessary for the model to be solvable by standard mathematical and statistical techniques. True linear relationships are probably very rare in the real world, yet linear-programming models and optimization techniques are based on an assumption of linearity. Similarly, many waiting-line systems probably violate some of the assumptions made in deriving the analytic queuing formulas used in Chapter 11. Even with these simplifications, however, there are many occasions where an analytic model will provide a meaningful, if not an excellent, solution to a problem.

By contrast, simulation models are not designed to be solved as such but to be experimented with in order to learn the probable effects of a decision. Since the intent is not to arrive at a mathematical solution, simulation models can accommodate very complex and realistic conditions. The general approach to conducting a simulation analysis is to develop a model that realistically portrays the situation and then to ask a series of "what if" questions to see what happens. In this manner, the consequences of alternative policies or operating procedures can be learned from the model. This experimentation is an iterative process that typically requires a large number of trials. Because replication plays such an important role, simulation is often referred to as a numeric technique.

Simulation is a versatile technique because the mathematical model can be made as simple or complex as necessary to describe the phenomenon under study. For example, a simulation model to describe the likely effect of tossing a die would be rather simple to construct. On the other hand, most macroeconomic simulation models are very complicated as a result of the many cause-and-effect relationships operating in the economy. The simulation technique can be applied in either case.

An important practical limitation on the use of simulation is that it tends to be a relatively expensive method of analysis. This is because many trials must be generated in order to observe the behavior of the simulated system under a variety of conditions. For that reason, most practical simulations require the use of a computer. It is not surprising that the growth and popularity of simulation has paralleled the development of computer technology.

The distinction between analytic and simulation models is important, because a decision maker might very well have to decide which type of model is appropriate for a given situation. An analytic model typically oversimplifies the problem but generally provides a direct solution. A simulation model usually provides a fairly realistic representation of the problem but can be analyzed only through tedious replication and experimentation. While there definitely exists a trade-off, deciding which approach to take is not as difficult as it might seem. Even the most zealous advocates of simulation would agree with the following statement: *In any situation where either an analytic approach or a simulation approach would be appropriate, the decision maker should always use the analytic approach.*

This statement simply means that if a problem situation is reasonably compatible with the assumptions of an analytic technique, the analytic approach should be used. The result will typically be a straightforward analysis that yields an excellent solution (and one that is probably not at all obvious). There are countless examples of the successful application of analytic techniques to management problems, and much of this book is concerned with describing and illustrating the most frequently used analytic models.

What simulation provides is an alternative for those situations where the assumptions required by an analytic technique are simply too restrictive or too unrealistic for the problem under study. For that reason, simulation has been referred to by some as a method of last resort. Nonetheless, it is a powerful technique that has been successfully applied to many important decision situations.

THE BASICS OF MONTE CARLO SIMULATION

While there are various types of simulation models, we focus our attention on an important category of models known as Monte Carlo simulation. This term refers to the idea of sampling at random from a probability distribution that describes some real-life phenomenon. By repeating the sampling procedure over and over, it is possible to simulate the occurrence of the phenomenon as it actually happens in the real world. The name "Monte Carlo" is descriptive of the chance or random nature of the process.

The basic idea behind Monte Carlo simulation is simple. To illustrate the procedure we shall develop an example. Suppose that one of the most profitable items carried by a large wholesale outlet is a particular brand of coffee maker. Weekly orders for the coffee maker have been recorded for the last 100 weeks, and the data are summarized in Table 12-1. The data indicate that between 9 and 15 coffee makers are demanded each week, with the distribution as shown.

TABLE 12-1   Weekly demand

Demand per week	Frequency	Relative frequency
9	5	.05
10	10	.10
11	15	.15
12	40	.40
13	15	.15
14	10	.10
15	5	.05
	100	1.00

If the data in Table 12-1 were recorded over a period of normal operations, the relative frequencies can be treated as probabilities. In deciding how many coffee makers to carry in stock, the wholesaler must somehow estimate how many coffee makers will be needed to fill orders over a given time. We develop an inventory-simulation model for this example later, but let us now simply illustrate how the Monte Carlo technique works by simulating the weekly orders for coffee makers.

From Table 12-1 the probability that only 9 coffee makers will be ordered in a given week is .05. Similarly, the probability that orders for 10 coffee makers will be received is .10. From week to week, the orders would be expected to vary from 9 to 15 according to the probabilities shown. One way to simulate the weekly orders is to take some paper and cut it up into 100 squares of equal size. On 5 of the 100 paper squares, the words "Demand equals 9" could be written. Similarly, 10 of the 100 paper squares would indicate that "Demand equals 10." Since 15 percent of the time 11 coffee makers are ordered during the week, 15 of the 100 paper squares would be designated "Demand equals 11." In similar fashion, weekly orders for 12, 13, 14, and 15 coffee makers would be indicated on 40, 15, 10, and 5 of the paper squares, respectively.

The 100 paper squares, each representing a value for weekly demand, could then be put into a box and mixed together thoroughly. To simulate the number of coffee makers ordered in a given week, one would simply reach into the box and randomly select a paper square. The demand indicated would represent simulated orders for that week. By repeating the process over and over (each time replacing the previously selected paper square), a distribution of weekly demands would be generated that exhibits the statistical properties recorded in Table 12-1.

By cutting up and labeling the 100 paper squares, an artificial process is created that behaves essentially like the real-world process. In the terminology of simulation, the box full of labeled pieces of paper would be called a *process generator*. The term refers to any artificial mechanism used to duplicate or imitate a real-world phenomenon. Since only 5 of the 100 pieces of paper indicate a demand of 9 coffee makers, there is only a 5 percent chance that a paper square labeled "Demand equals 9" will be drawn from the box. That, of course, is exactly the probability that 9 coffee makers will be ordered in any given week in the real world. Thus, the artificial process of randomly drawing from the box is used to represent the real process of weekly demand.

The process generator is an extremely important part of any Monte Carlo simulation model. In the real world, it takes an entire week to generate one value for the variable "weekly demand." With the process generator, weekly demand occurs whenever we want it to. This allows the modeler to advance the clock and simulate the number of orders that will be received several periods into the future. As we shall demonstrate, this ability is one of the unique features of simulation analysis.

The Use of Random Numbers

Fortunately, it is not necessary to cut up and label paper squares every time a process generator is needed in a simulation model. Instead, a sequence of random numbers is used repeatedly to sample from the probability distribution of interest. A random-number sequence is simply a series of numbers in which each possible number has an equal chance of occurring. For example, a list of one-digit numbers would be a random number sequence if every digit, zero through nine, had an equal chance of occurring when a number was added to the list. This is analogous to the 100 paper squares in the box that each have an equal chance of being drawn. The simulation is conducted by letting each random number represent a particular event that could occur.

Random-number sequences can easily be generated by a computer. This is especially important because many practical simulations are so large that a computer must be used. In fact, most computer systems are capable of generating random numbers one at a time, as they are needed in the simulation model. For simulations that are to be conducted manually (as we shall do here), the computer generates an entire series of random numbers at once and prints out the sequence. One such sequence is presented in Appendix H.

Every digit in the random-number table in Appendix H was generated by a random process. Even though the numbers are grouped in clusters of five digits, every digit is a random number. Thus, any way that you systematically move through the table (up or down the columns, across the rows, even diagonally) a random-number sequence is generated. For example, taking the first two digits of the five-digit cluster and moving down the column, a series of two-digit random numbers between 00 and 99 is generated. Since we shall be using random numbers for several illustrations in this chapter, part of the two-digit sequence just described is recorded in Table 12-2 for reference.

TABLE   12-2   Two-digit random numbers

10	98	85	02
37	11	16	85
08	83	65	01
99	88	76	97
12	99	93	63
66	65	99	52
31	80	65	53
85	74	70	11
63	69	58	62
73	09	44	28

In order to use the random numbers to simulate weekly orders for the coffee maker, we must associate with each random number some level of demand. Since we are working with two-digit random numbers, there are 100 possible numbers that appear (00 through 99). Each number in the interval corresponds to a paper square in the previous example, and all 100 numbers have an equal chance of occurring. Just as we wanted 5 of the 100 paper squares to represent a weekly demand of 9 coffee makers, we want 5 of the 100 random numbers to be associated with a demand of 9. Thus, we could use random number 00 through 04, inclusive, to represent a weekly demand of 9. In similar fashion, 10 of the 100 random numbers (05 to 14) will represent weekly orders for 10 coffee makers. Assigning the random numbers in this fashion yields the process generator in Table 12-3.

TABLE 12-3   Process generator for weekly demand

Weekly demand	Probability	Random numbers
9	.05	00–04
10	.10	05–14
11	.15	15–29
12	.40	30–69
13	.15	70–84
14	.10	85–94
15	.05	95–99

The simulation is conducted by systematically taking a random number from Table 12-2 and associating it with the corresponding level of demand in the process generator. Simulated orders for a 6-week period are shown in Table 12-4. The number of coffee makers ordered over the simulated period varies according to the probabilities specified. If we were to continue the simulation over a much longer period and summarize the resulting data, we would expect to duplicate the probability distribution for weekly demand in Table 12-1. Generally, long simulation runs are necessary to generate simulated results that correspond in a statistical sense to the real-world phenomenon under study. That is one reason why simulation is an expensive and time-consuming method of analysis.

TABLE 12-4   Simulated orders for a 6-week period

Simulated week	Random number	Simulated demand
1	10	10
2	37	12
3	08	10
4	99	15
5	12	10
6	66	12

Two or More Probability Distributions

We have illustrated how the Monte Carlo technique works in sampling from a single probability distribution. In many decision situations a number of random variables (each described by some probability distribution) must be considered in order to evaluate alternative courses of action. Whenever two or more random variables are related, uncertainty and variability can cause strange things to happen. For example, it is often chance interactions between random variables that lead to such things as production delays, inventory stock-outs, project overruns, and so forth.

Simulation provides an opportunity to see the likely effects of chance variations when two or more random variables are related. In effect, a model is an explicit statement of how the variables are related, and the Monte Carlo technique is a means of sampling from the probability distribution that describes each random variable. A representative simulation run is likely to provide important insights into the behavior of the system.

In the remainder of this chapter we develop and illustrate three simulation models. First we introduce some additional data concerning the coffee-maker example and develop a simulation model to analyze a possible inventory policy for the item. Then the application of simulation to a waiting-line problem and a project-scheduling problem will be illustrated.

INVENTORY SIMULATION

Let us assume that the following data pertain to the coffee maker in the previous example. The wholesaler buys the coffee makers directly from the manufacturer at a cost of $20 each and sells them to retailers at $26 each. The wholesaler incurs a $50 cost each time an order is placed with the manufacturer and estimates that each coffee maker carried in inventory for 1 year creates $5.20 in cost.

The inventory problem faced by the wholesaler is twofold: (1) How many coffee makers should be ordered from the manufacturer and carried in stock? (2) When should reorders be placed? These are the typical questions of quantity and timing that are important in developing any inventory policy. What makes this situation more complicated is the fact that the wholesaler does not know how many coffee makers will be needed in any given week to meet demand from the retailers. The variability in weekly demand was illustrated in Table 12-1. Let us make the situation even more complicated and realistic by assuming that the wholesaler also does not know how long it will take for a reorder to arrive from the man-

TABLE 12-5   Lead time

Lead time from manufacturer to wholesaler, weeks	Frequency	Relative frequency	Random numbers
1	4	.20	00–19
2	12	.60	20–79
3	4	.20	80–99
	20	1.00	

ufacturer. Analysis of the past 20 reorder cycles for the coffee maker indicates that lead time varies between 1 and 3 weeks. These data are summarized in Table 12-5. Also shown are random numbers corresponding to the probabilities.

The wholesaler's inventory problem is illustrated by Figure 12-1. The Monte Carlo simulation model provides a means of explicitly addressing the uncertainty expressed in weekly demand and lead time. The interaction between these two random variables can have a significant effect on the costs of operating the inventory system. For example, each time an order is placed, there is uncertainty about how long it will take the manufacturer to make delivery *and* how many coffee makers will be demanded at the wholesale outlet during the period. The combination of unusually long delivery periods and unusually heavy demand could result in shortages and lost sales for the wholesaler. On the other hand, the combination of unusually short delivery periods and unusually light demand could leave the wholesaler with excessive inventory and high carrying costs.

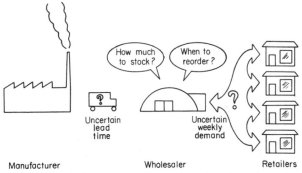

Figure 12-1   The wholesaler's inventory problem.

Since demand and lead time are expressed on a weekly basis, we shall develop a model to simulate the weekly operation of the inventory system. The following notation is used to represent the inventory costs associated with the coffee maker:

$$\text{Ordering cost } C_O = \$50 \text{ per order}$$
$$\text{Weekly carrying cost } C_K = \frac{\$5.20}{52} = \$.10 \text{ per unit per week}$$
$$\text{Stock-out cost } C_S = \$26 - \$20 = \$6 \text{ per unit}$$

The weekly carrying cost of $.10 per unit is computed by simply dividing the estimated annual carrying cost of $5.20 by the number of weeks in the year (52). The stock-out cost of $6 represents the difference between the wholesaler's selling price and cost. This implies that no backlog is kept and orders that the wholesaler cannot fill from inventory are lost. We shall assume that this is the case in order to keep the illustration as simple as possible, although the simulation model is capable of handling more complex situations as well.

Assuming that the wholesaler is concerned with the effects of the inventory policy on operating costs, the model can be expressed as

$$\begin{matrix}\text{Weekly inventory} \\ \text{cost of} \\ \text{coffee maker}\end{matrix} = \begin{matrix}\text{weekly} \\ \text{stock-out} \\ \text{cost}\end{matrix} + \begin{matrix}\text{weekly} \\ \text{carrying} \\ \text{cost}\end{matrix} + \begin{matrix}\text{weekly} \\ \text{ordering} \\ \text{cost}\end{matrix}$$

or

$$\begin{matrix}\text{Weekly inventory} \\ \text{cost of} \\ \text{coffee maker}\end{matrix} =$$

$$\begin{bmatrix}\text{number of} \\ \text{coffee makers} \\ \text{demanded but} \\ \text{not in stock} \\ \text{during week}\end{bmatrix}(C_S) + \begin{bmatrix}\text{number of} \\ \text{coffee makers} \\ \text{carried in} \\ \text{stock through-} \\ \text{out week}\end{bmatrix}(C_K) + \begin{bmatrix}\text{number of} \\ \text{orders placed} \\ \text{with} \\ \text{manufacturer} \\ \text{during} \\ \text{week (0 or 1)}\end{bmatrix}(C_O)$$

In terms of the example problem, the model becomes

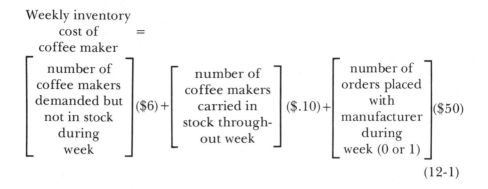

$$(12\text{-}1)$$

For each simulated week, a numerical value will be generated for each variable in brackets in formula (12-1), and the weekly cost will be computed. The process will be repeated over and over to develop an estimate of the average weekly cost. The values generated for the variables in brackets will be affected by two factors. The first is the inventory policy established by the wholesaler, that is, the order quantity and reorder point for the coffee maker. The second factor is the chance interaction between lead time from the manufacturer and weekly demand at the warehouse. While the inventory policy can be controlled by the wholesaler, the uncertainty and variability in demand and delivery time cannot.

Let us assume that the wholesaler is considering implementing an inventory policy of ordering coffee makers from the manufacturer in lots of 100. An order would be placed when the wholesaler's stock level falls to 25. This is the inventory policy we want to simulate in order to estimate average weekly cost. The policy to be simulated can be expressed as

$$\text{Order quantity } Q = 100$$
$$\text{Reorder point } R = 25$$

The process generators developed in Table 12-3 for weekly demand and Table 12-5 for delivery time will be used to generate the random variables in the model. For random numbers, we shall use the two-digit sequence reproduced in Table 12-2, beginning at the left-hand column and moving downward. For each simulated week, then, a random number will be used to simulate orders for the coffee maker, just as illustrated in the previous example. When the simulated stock level falls to 25, a

TABLE 12-6 Inventory simulation

Simulated week	Received from previous order	Beginning balance	Random number for demand	Simulated demand	Sales	Demanded but not in stock*	Ending balance*	Reorder?* †	Random Number for lead time	Simulated lead time	Stock-out cost	Carrying cost	Ordering cost	TC
													Weekly cost [Formula (12-1)]	
1		100	10	10	10	0	90	0			$ 0	$9.00	$ 0	$ 9.00
2		90	37	12	12	0	78	0			0	7.80	0	7.80
3		78	08	10	10	0	68	0			0	6.80	0	6.80
4		68	99	15	15	0	53	0			0	5.30	0	5.30
5		53	12	10	10	0	43	0			0	4.30	0	4.30
6		43	66	12	12	0	31	0			0	3.10	0	3.10
7		31	31	12	12	0	19	1	85		0	1.90	50	51.90
8		19	63	12	12	0	7	0		3	0	.70	0	.70
9		7	73	13	7	6	0	0			36	0	0	36.00
10		0	98	15	0	15	0	0			90	0	0	90.00
11	100	100	11	10	10	0	90	0			0	9.00	0	9.00
12		90	83	13	13	0	77	0			0	7.70	0	7.70
13		77	88	14	14	0	63	0			0	6.30	0	6.30
14		63	99	15	15	0	48	0			0	4.80	0	4.80
15		48	65	12	12	0	36	0			0	3.60	0	3.60
16		36	80	13	13	0	23	1	74	2	0	2.30	50	52.30
17		23	69	12	12	0	11	0			0	1.10	0	1.10
18		11	09	10	10	0	1	0			0	.10	0	.10
19	100	101	85	14	14	0	87	0			0	8.70	0	8.70
20		87	16	11	11	0	76	0			0	7.60	0	7.60
							901							316.10

*Variables needed in formula (12-1).

† 1 = yes; 0 = no.

random number will be used to simulate the time required for the manufacturer to deliver the order for 100 coffee makers.

Record keeping is an important part of the simulation analysis since for each simulated week we must determine (1) the number of coffee makers demanded but not in stock, (2) the number of coffee makers carried in inventory, and (3) whether or not an order had to be placed. These are the variables in brackets in formula (12-1) that are used to compute weekly inventory cost. Table 12-6 illustrates how the information from the simulation is recorded and shows the results of a 20-week simulation run.

Each row in Table 12-6 represents a simulated week, and each column heading represents a piece of information generated from the model. The three column headings marked by an asterisk represent the variables needed to make the weekly cost calculation in formula (12-1). Note that week 1 starts out with the entire order quantity (100) in stock. The question of how to start a simulation model of this type is an interesting and important one that we come back to later.

Week 1 is simulated by comparing the first random number in Table 12-2 (10) with the demand process generator (Table 12-3). This generates simulated orders for 10 coffee makers, which are easily filled from the wholesaler's inventory. The stock level falls to 90 (100 − 10); no reorder is needed in week 1, nor has any stock-out cost been incurred. The weekly cost computed for the first simulated week is

$$0(\$6) + 90(\$.10) + 0(\$50) = \$9$$

Simulated weeks 2 through 6 show a fluctuating weekly demand, but no stock-outs occur and the inventory level remains above the reorder point. In week 7, orders for 12 coffee makers reduce the stock level below 25, and a replenishment order of 100 must be placed with the manufacturer. To simulate the delivery period, the next random number in the sequence (85) is compared with the lead-time process generator (Table 12-5) to yield a 3-week delivery period. Thus, the replenishment order placed at the end of week 7 will be received at the end of week 10 and is therefore available to fill retail orders in week 11. This is recorded by week 11 in the table. The weekly inventory cost in week 7 includes the $50 ordering cost.

Simulated demand during the 3-week delivery period is greater than the remaining stock, and stock-outs are recorded in weeks 9 and 10 before the order arrives. The inventory cost for these 2 weeks reflects the opportunity cost of the lost sales. Upon receipt of the order from the manufacturer, the stock level increases in week 11 and current demand is met from inventory.

As the simulation continues, the inventory is again reduced to its reorder level in week 16, and a replenishment order of 100 coffee makers must be placed. This time, however, the delivery period is only 2 weeks, and the remaining stock is sufficient to meet demand until the order arrives. Thus, no stock-outs are experienced during the second reorder cycle. At the end of the 20-week simulated period, there are 76 coffee makers in the wholesaler's inventory.

Summary Statistics

At the end of a simulation run it is necessary to analyze the results and summarize what has occurred. In this case, the primary concern is to estimate the average weekly cost generated by a given inventory policy ($Q = 100$, $R = 25$) when weekly demand and lead time are subject to uncertainty and variability. Based on the simulation just completed, the estimated cost is found by computing the average of the 20 weekly cost figures generated by the model. Thus, our best estimate of weekly inventory cost under the proposed policy is

$$\text{Average weekly cost} = \frac{\Sigma \text{ simulated weekly costs}}{\text{no. of simulated weeks}} = \frac{\$316.10}{20} = \$15.81$$

The data in Table 12-6 also provide other relevant information about the proposed inventory policy. For example, an estimate of the average number of coffee makers in the wholesaler's inventory can be computed by averaging the 20 ending-balance figures. The calculation yields

$$\text{Average inventory} = \frac{\Sigma \text{ simulated ending balances}}{\text{no. of simulated weeks}} = \frac{901}{20} = 45.05$$

Also, it might be important to know the probability that a stock-out condition will occur each time an order is placed with the manufacturer. That is, how likely is it that 25 coffee makers will not be enough to fill retail orders until the manufacturer's shipment arrives? In the 20-week simulation, orders were placed in week 7 and week 16. The order placed in week 7 took 3 weeks for delivery, and the remaining stock was not sufficient to meet demand until the order arrived. A total demand for 21 coffee makers could not be met because of the stock-out. On the other hand, the order placed in week 16 arrived 2 weeks later, and no sales were lost during the lead time.

From the data developed in the simulation, the estimated probability of incurring a stock-out in any given reorder cycle would be .50. However, this would not seem to be a very reliable estimate, since it is based on only two observations. In general, the reliability of any summary statistic computed from a simulation run increases with the number of observations. This simply means that long simulation runs are likely to provide much more meaningful results than short simulation runs. While the 20-week exercise served to illustrate how an inventory simulation model works, it probably is not of sufficient length to permit meaningful conclusions to be drawn about the inventory policy being simulated.

Length of Run and Starting Conditions

The importance of the length of a simulation run becomes clear when simulation is viewed as a sampling process. Each simulated period adds to the sample of data from which a parameter estimate is to be made. In the inventory simulation, the parameter of primary interest is the average weekly cost generated by a particular ordering policy. The estimate is made from a sample of weekly observations generated by the simulation model. It stands to reason that a large sample is more likely than a small one to exhibit the same characteristics as the population.

It should not be surprising that many important concepts in statistical sampling theory also apply to the design of simulation experiments. In fact, it is possible to compute the number of simulated observations necessary to have a specified level of confidence in a resulting parameter estimate. This is nothing more than a standard sample-size calculation. We do not illustrate that type of analysis here but emphasize the importance of generating representative simulation runs.

Fortunately, it is rather easy to increase the sample size in most simulation models. The difficulty and expense occur largely in designing the model. Once the model is operational, it is an easy matter to continue generating observations. This, of course, is particularly true when a computer is used in the analysis.

Another important consideration relating to sample size is how the model is started. In some simulation models, the starting point must be arbitrarily chosen. This was essentially the case in the inventory simulation, where it was assumed that the entire order quantity ($Q = 100$) constituted the beginning inventory for the first simulated week. Since the modeler selected the initial conditions, it is difficult to know what effect the starting point had on the final results.

In general, two things can be done to offset the problem associated with arbitrarily selected starting conditions. The first is to make sure that the simulation run is a long one. The longer the simulation run the less important the initial conditions. In fact, if the simulation is run long enough, the process being simulated will find its "natural equilibrium" regardless of the initial conditions. The second approach is to let the model itself generate the starting conditions. This can be done by starting at some arbitrarily selected point and generating a series of observations. After a number of simulated periods have been generated, the model is stopped and the status of the system is used as the starting point for beginning the actual simulation. This is equivalent to running the model for a period without keeping score in order to achieve a more realistic starting point.

Comparing Alternative Policies

Since simulation models are descriptive, their purpose is to determine what might happen if a particular strategy or operating policy is followed. The inventory simulation just illustrated was designed to show how the inventory system would operate under a policy of ordering 100 coffee makers when the stock level falls to 25. The model does not indicate whether this is the optimal policy; it only shows what is likely to happen if it is put into effect. To evaluate alternative inventory policies, one would repeat the simulation process, considering the alternative order quantities and reorder points. A comparison of the relevant summary statistics for each simulation provides a good indication of which policy would have the desired effects.

WAITING-LINE SIMULATION

Suppose a bank is planning to open a branch office for which it must decide whether or not to have a drive-in window. In order to gain some insight into the operation of the proposed drive-in window, it is decided to construct a simulation model. Of major concern to management is the average time it takes to go through the waiting line and complete a transaction.

The first step in the construction of the model is to describe the nature of the arrival and service processes. Since both the arrival of customers and the time necessary to serve them are random variables, a probability distribution is developed to describe each process. The expected arrival of customers is expressed in the form of a distribution of interarrival times, that is, the time that elapses between successive arrivals into the system. In a simulation model, it is more convenient to describe the arrival process in the form of an *interarrival-time* distribution than an *arrival-rate*

TABLE 12-7 *Distribution of interarrival times*

Interarrival time, minutes	Probability	Random numbers
3	.10	00–09
4	.15	10–24
5	.25	25–49
6	.20	50–69
7	.15	70–84
8	.10	85–94
9	.05	95–99
	1.00	

distribution, as we did in Chapter 11. Table 12-7 shows how the time between arrivals at the proposed drive-in window is expected to vary. Also shown are random numbers associated with each interarrival time for use as the process generator.

The service time at the drive-in window is expected to vary between 1 and 7 minutes according to the probabilities in Table 12-8. The variations in service time reflect the fact that some banking transactions take longer to complete than others. Random numbers are assigned to each service time in proportion to the probabilities given. It should be noted that the distribution of service times in Table 12-8 does *not* take the shape of an exponential distribution and therefore the analytic queuing formulas illustrated in Chapter 11 do not apply.

The average time required to go through the system will depend greatly on the chance interactions between the interarrival- and service-time distributions. For example, if interarrival times are short and service times are long, there is likely to be excessively long waiting lines at the

TABLE 12-8 *Distribution of service times*

Service time, minutes	Probability	Random numbers
1	.05	00–04
2	.10	05–14
3	.20	15–34
4	.30	35–64
5	.20	65–84
6	.10	85–94
7	.05	95–99
	1.00	

drive-in window. On the other hand, the combination of long interarrival times and short service times would mean that cars could go through the system very quickly, and the teller at the drive-in window would be idle much of the time.

In order to model the process, we shall generate simulated arrivals at the drive-in window and keep track of the time it takes each to go through the system. By repeating the process over and over, we should be able to develop a reliable estimate of the average time it will take for a car to be processed at the window. The simulation will start at time zero, and we shall record the cumulative time at which each simulated arrival (1) enters the system, (2) begins being served by the teller, and (3) completes the transaction and leaves the system. For each arrival the following statistics will be computed:

$$\text{Waiting time} = \text{time service began} - \text{time of arrival} \qquad (12\text{-}2)$$

and

$$\text{Time in system} = \text{time service finished} - \text{time of arrival} \qquad (12\text{-}3)$$

The time of each simulated arrival will be generated from Table 12-7 (which yields the elapsed time from the previous arrival), and the simulated service time for each arrival will come from the process generator in Table 12-8. Again, we shall use the random numbers in Table 12-2, beginning at the top of the left-hand column and moving downward.

Table 12-9 shows the results of a simulation of 20 arrivals at the drive-in window. For customer 1, the first random number (10) generated an interarrival time of 4 minutes, indicating that the first customer arrived 4 minutes after the simulation began. Since we started the simulation with the system "empty," customer 1 experienced no waiting time and was able to drive right up to the window at time 4. The next random number is 37, which generates a service time of 4 minutes for customer 1. Thus, the first simulated customer arrived at time 4, experienced no waiting line, was served and left the system at time 8. The relevant statistics for customer 1 are shown in the last two columns.

Customer 2 has an interarrival time of 3 minutes. This indicates that customer 2 arrived at time 7 (3 minutes after the arrival of customer 1). Since the first customer was in the system until time 8, the arrival at time 7 had to wait 1 minute before the teller at the drive-in window was free. Thus, customer 2 began service at time 8, when the service facility became available. In general, the time that service begins for a simulated customer is the larger of (1) the time of arrival for that customer and (2) the time

TABLE 12-9 Waiting-line simulation

| Simulated customer | Simulated arrivals | | | Have to wait? | Time service began* | Simulated Service | | | Waiting time [formula (12-2)] | Time in system [formula (12-3)] |
	Random number	Interarrival time	Time of arrival*			Random number	Service time	Time service finished*		
1	10	4	4	no	4	37	4	8	0	4
2	08	3	7	yes	8	99	7	15	1	8
3	12	4	11	yes	15	66	5	20	4	9
4	31	5	16	yes	20	85	6	26	4	10
5	63	6	22	yes	26	73	5	31	4	9
6	98	9	31	no	31	11	2	33	0	2
7	83	7	38	no	38	88	6	44	0	6
8	99	9	47	no	47	65	5	52	0	5
9	80	7	54	no	54	74	5	59	0	5
10	69	6	60	no	60	09	2	62	0	2
11	85	8	68	no	68	16	3	71	0	3
12	65	6	74	no	74	76	5	79	0	5
13	93	8	82	no	82	99	7	89	0	7
14	65	6	88	yes	89	70	5	94	1	6
15	58	6	94	no	94	44	4	98	0	4
16	02	3	97	yes	98	85	6	104	1	7
17	01	3	100	yes	104	97	7	111	4	11
18	63	6	106	yes	111	52	4	115	5	9
19	53	6	112	yes	115	11	2	117	3	5
20	62	6	118	no	118	28	3	121	0	3
									27	120

*Variables needed in formulas (12-2) and (12-3).

service was finished for the previous customer. The simulated service time was 7 minutes for customer 2, who left the system at time 15.

As the simulation continues, customer 3 arrived at time 11 but had to wait in line 4 minutes until the teller was available at time 15. In similar fashion, customers 4 and 5 also had to wait for the teller to finish with the previous arrival. The relatively long period between the arrivals of customers 5 and 6 enabled the teller to "catch up," and customers 6 through 13 were served immediately upon arrival. The waiting line developed again later as customers 16 through 19 had to endure varying amounts of waiting time before the teller was available. At the end of the simulation run, 20 customers have been generated covering approximately a 2-hour period (121 simulated minutes).

Summary Statistics

While our previous comments about the desirability of long simulation runs also apply in this case, we shall summarize the data generated from the 20-customer simulation. An estimate of the average waiting time at the drive-in window is developed by averaging the 20 waiting times generated by the model

$$\text{Average waiting time} = \frac{\Sigma \text{ waiting times}}{\text{no. of simulated customers}}$$
$$= \frac{27}{20} = 1.35 \text{ minutes}$$

In similar fashion, the average time in the system is estimated as

$$\text{Average time in system} = \frac{\Sigma \text{ times in system}}{\text{no. of simulated customers}}$$
$$= \frac{120}{20} = 6 \text{ minutes}$$

Finally, we can estimate from the data generated in the simulation the probability that an arrival will have to wait. This estimate is simply the proportion of the 20 simulated customers who experienced a waiting line at the drive-in window. This figure also represents the proportion of time the teller in the drive-in window is expected to be busy. The computation yields

$$\text{Probability arrival will have to wait} = \frac{\text{no. of customers having to wait}}{\text{no. of simulated customers}}$$
$$= \frac{9}{20} = .45$$

PERT SIMULATION

As a final illustration of the application of simulation, let us consider a project to install some new equipment in a production facility. Assume that the PERT network in Figure 12-2 portrays the project activities and their relationships. We want to simulate the completion of the activities in order to estimate how long it will take to finish the project.

Table 12-10 describes in the form of probability distributions the uncertainty that exists about how long it will take to complete each activity. Also shown in the table are random numbers associated with each possible completion time. In the analytic PERT model in Chapter 9, the expected value of each activity-time distribution was used as an estimate of the duration of the activity. From the expected values indicated in Table 12-10, the critical path through the network in Figure 12-2 would consist of activities B-C-D (events 1-2-4-5-6), and the analytic estimate of the project length would be 17 days.

Simulation analysis provides a more realistic method of estimating the project duration by allowing the completion of each component activity to vary according to the probabilities given. The simulation consists of generating at random a completion time from each activity distribution and then determining what the critical path would be and how long the project would take if those activity times actually did occur. By repeating the process many times, a distribution of project durations is generated from which the average duration can be computed.

The two-digit random numbers in Table 12-2 are used to sample from the four activity distributions, and the results of 10 iterations are summarized in Table 12-11. In each iteration, a completion time is generated for each activity, and the critical path and project duration are determined from the network. In the first iteration, the two possible paths were equal in length, and the project would have been finished in 16 days had those completion times actually occurred.

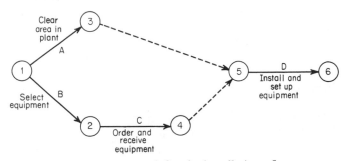

Figure 12-2 PERT network for the installation of new equipment.

TABLE 12-10   Data for activities A to D

Completion time, days	Probability	Random numbers	Expected value, days
Activity A, clear area in plant			
8	.10	00–09	
9	.20	10–29	
10	.40	30–69	
11	.20	70–89	
12	.10	90–99	
	1.00		10
Activity B, select equipment			
2	.25	00–24	
3	.50	25–74	
4	.25	75–99	
	1.00		3
Activity C, order and receive equipment			
6	.20	00–19	
7	.20	20–39	
8	.20	40–59	
9	.20	60–79	
10	.20	80–99	
	1.00		8
Activity D, install and set up equipment			
5	.25	00–24	
6	.50	25–74	
7	.25	75–99	
	1.00		6

In iteration 2, activities B-C-D are critical, and the project duration would be 17 days. As the simulation continues, the project duration varies between a high of 20 days in iteration 4 and the low of 16 days in the first iteration. The 17-day estimate derived from the analytic PERT procedure holds true in only 4 of the 10 iterations.

Summary Statistics

The estimated length of the project derived from the simulation analysis is determined by averaging the 10 project durations. It should be noted that the starting conditions in this model do not bias the estimate in any way, since each iteration is completely independent of the others. Based on the 10 iterations, the estimated project duration is

TABLE 12-11 PERT simulation

Iteration	Activity	Random number	Simulated completion time	Paths		Project duration, days
				A-D	B-C-D	
1	A	10	9	16*	16*	16
	B	37	3			
	C	08	6			
	D	99	7			
2	A	12	9	16	17*	17
	B	66	3			
	C	31	7			
	D	85	7			
3	A	63	10	15	18*	18
	B	73	3			
	C	98	10			
	D	11	5			
4	A	83	11	17	20*	20
	B	88	4			
	C	99	10			
	D	65	6			
5	A	80	11	16	17*	17
	B	74	3			
	C	69	9			
	D	09	5			
6	A	85	11	18*	18*	18
	B	16	2			
	C	65	9			
	D	76	7			
7	A	93	12	18	19*	19
	B	99	4			
	C	65	9			
	D	70	6			
8	A	58	10	17*	16	17
	B	44	3			
	C	02	6			
	D	85	7			
9	A	01	8	14	19*	19
	B	97	4			
	C	63	9			
	D	52	6			
10	A	53	10	16	17*	17
	B	11	2			
	C	62	9			
	D	28	6			
						178

*Critical path.

$$\text{Estimated project duration} = \frac{\Sigma \text{ project durations}}{\text{no. of project simulations}}$$

$$= \frac{178}{10} = 17.8 \text{ days}$$

Another important piece of information gained from the simulation analysis concerns the activities likely to be on the critical path. While the number of days required to finish the project varied considerably from iteration to iteration, the activities constituting the critical path remained almost unchanged. In 9 of the 10 iterations, the path consisting of activities B-C-D was critical to the project duration. This is strong evidence that close control over activities B, C, and D is necessary for the timely completion of the project.

CONCLUSION

Simulation is a method of analysis that can be used when the problem under study cannot be modeled adequately by an analytic technique. The versatility of simulation analysis was demonstrated in this chapter with three illustrative applications. While it is generally an expensive method to use, simulation often provides information that cannot be obtained by other means. In many cases, the alternative would be to experiment with the real system, a risky and potentially disastrous way to evaluate decision alternatives.

The growth and importance of simulation is reflected by the fact that several computer programming languages have been developed especially for use in simulation analysis. The languages are designed to be compatible with the special nature of simulation studies. Typically, they provide a convenient means to generate random numbers, clock and record simulated observations, and handle the highly repetitive nature of the analysis. Also, many of the programming "words" are similar to the technical terms in simulation, making it easier to translate the model into instructions understandable to the computer.

In recent years, simulation has received much attention in the professional and academic literature. It is also a popular topic in many management development seminars. The versatility of the technique and the growing body of computer software to support it make simulation analysis an important management tool.

PART 4
Decision-Making Applications

Although the statistical and quantitative methods presented in Parts 2 and 3 are, wherever possible, illustrated with managerial data, the situations are contrived to support the method. In practice, of course, a manager expects to function in exactly the opposite manner, selecting a method to satisfy a particular decision-making need.

In this part, we present a number of managerial decision-making situations in the form of cases that require the use of some of the statistical and quantitative methods described in Parts 2 and 3. Although we have tried to make these cases realistic, the reader should recognize certain artificialities. One major difference between the real world and the world of cases lies in the collection of data. Of necessity, all the data required to solve each case (and virtually no other data) are furnished. In the real world, the decision maker must identify the necessary data, screen out unwanted data, collect the data required, verify them, and convert them into the proper format for computation. Often, particularly where the actual computation is performed with a computer, data collection is the single most difficult task facing the decision maker.

Cases are meant to be solved *rather than* read. *The reader will derive the maximum benefit from each case by stopping at each decision step and answering the questions facing the characters in the case at that point. If, in some instances, this means stopping to perform a calculation or to look up a formula, by all means take the time to do it. If errors are made or a* path different from ours is taken, the preferred solution can easily be picked up before the next decision point. Formulas used in the solution of each case are identified by the same numbers used in Chapters 3 to 11 and Appendix C.*

The cases presented here are examples *of decision-making applications in various functional areas. They by no means represent an exhaustive inventory of the uses of statistics and quantitative methods in those areas. For such a treatment, we refer the reader to specialized texts in the areas of finance, accounting, production, marketing, and personnel administration.*

CHAPTER 13

Financial Management Application

THE O'LEARY CASE

Thomas Wellington III and Ryan O'Leary had been close friends from the day they first met at the Northside Elementary School in the Boston suburb of Brookline. They shared nearly perfect academic records, an enviable array of athletic accomplishments, and, at one point in junior high, the same girl friend. After high school, the two followed distinctly different but equally successful paths. Tom received an appointment to the United States Military Academy at West Point, where his final academic average was second only to that of Douglas MacArthur. He distinguished himself in combat and in politically sensitive assignments at the Pentagon and late in his career was rewarded with the post of supreme commander of NATO forces.

Ryan attended Harvard, and although he received only average grades, he was admitted to Harvard Law School. Upon graduation from law school, Ryan joined a prestigious New York law firm and later began to dabble in New York politics.

When the Republican National Committee, stung by a resounding defeat in the last presidential election and sensing a groundswell of public longing for strong national leadership, decided to approach Tom with the idea of running for president, it was only natural that they chose Ryan as an intermediary. Tom agreed to run on the condition that Ryan manage his campaign. The result should have been predictable to anyone who had known the pair back in Brookline. Formidable as individuals, as a team they were unbeatable. Tom was swept into office in a landslide, and in his first appointment named Ryan Secretary of State.

From the Desk of Bob Butcher

MEMORANDUM

TO: Nancy Simpson
SUBJECT: O'Leary Portfolio
DATE: Dec. 2, 1980

I would like you to plan a portfolio for Secretary of State Designate O'Leary. In doing so, please consider the following:
   a. The total investment cannot exceed $1,352,l00.
   b. Initially, draw from the attached list of investment opportunities.
   c. Note that Treasury Bonds are in limited supply.
   d. Do not invest more than 20 percent of the total in any one firm (not applicable to government securities).
   e. Do not invest more than 30 percent of the total in any one industry (not applicable to government securities).
   f. Do not invest more than 40 percent of the total in growth stocks.
   g. Purchase at least $3 of government securities for every $1 of foreign investment.
   h. Invest at least 40 percent in income stock.
   i. Maximize the expected return on investment.

Figure 13-1   Memorandum from Bob Butcher to Nancy Simpson.

While Ryan was eager to accept the Cabinet post, he had some concern over the not inconsiderable fortune he had accumulated in 30 years of successful law practice. It was with great care, therefore, that he selected the investment firm of Butcher, Baker, and Blackstone to manage his blind trust while he was in office.

Bob Butcher, the senior partner, decided to manage the trust himself with assistance from Nancy Simpson, a recently hired Boston University MBA graduate.

To keep the trust blind, Bob felt that he had to liquidate all of Ryan's present holdings and to reinvest the proceeds in firms and institutions unknown to Ryan. Fortunately, the Republican sweep had stimulated the business sector, and Bob realized $1,352,100 from Ryan's assets. Bob planned to turn over the actual reinvestment to Nancy but felt that he should give her some guidance based on his knowledge of Ryan's investment philosophy. His memo to her on this subject is shown in Figure 13-1. Bob also drew up a list of investment opportunities and their expected aftertax yields from a company investment guide prepared monthly by the research staff. His list is shown in Figure 13-2.

Investment	Availability	Expected Annual Yield, %
U.S. Treasury Bonds:		
February 1993	150,000	7.74
August 1992	90,000	8.03
May 1990	60,000	8.32
Growth stock:		
Computer Technology Corp.	No limit	15.0
Goode's Discount Stores, Inc.	,,	18.0
Persian Gulf Oil (foreign)	,,	20.0
Yakamura Electronics (foreign)	,,	16.0
Income stock:		
National Transistor Corp.	,,	12.0
Oklahoma Oil Company	,,	10.0
Family Catalog Sales	,,	9.0
CONFIDENTIAL		

Figure 13-2 Investment opportunities: O'Leary portfolio.

Upon receiving Bob's note, Nancy considered the quantitative methods applicable to a portfolio selection problem. The instructions to maximize the expected return on investment suggested an *optimizing* model. Furthermore, she reasoned that return on investment was simply the sum of the amount invested in each stock or bond times its respective expected yield, a relationship that could be expressed *analytically* by a *linear equation*. Finally, she saw that the remaining guidelines also could be written as linear equations that would constrain the use of Mr. O'Leary's funds. Everything pointed to a *linear-programming* analysis, and she quickly redesignated the stocks and bonds suggested by Bob as indexed variables, as shown in Table 13-1.

Next, Nancy had to formulate the objective function and constraint equations to maximize return on investment within the guidelines established by Bob. After several attempts, she finally decided that the equations in Table 13-2 satisfied all explicit and implicit conditions. Before

TABLE 13-1   Designation of variables

Amount invested in	Variable	Amount invested in	Variable
Treasury Bond, February 1993	X_1	Persian Gulf Oil	X_6
Treasury Bond, August 1992	X_2	Yakamura Electronics	X_7
Treasury Bond, May 1990	X_3	National Transistor Corp.	X_8
Computer Technology Corp.	X_4	Oklahoma Oil Company	X_9
Goode's Discount Stores	X_5	Family Catalog Sales	X_{10}

TABLE 13-2   Linear-programming formulation*

Constraints:	
Fund availability:	$X_1 + X_2 + X_3 + X_4 + X_5 + X_6 + X_7 + X_8 + X_9 +$ $X_{10} \leq 1,352,100$
Bond availability:	$X_1 \leq 150,000$
	$X_2 \leq 90,000$
	$X_3 \leq 60,000$
Firm maximums:	$X_4 \leq 270,420$
	$X_5 \leq 270,420$
	$X_6 \leq 270,420$
	$X_7 \leq 270,420$
	$X_8 \leq 270,420$
	$X_9 \leq 270,420$
	$X_{10} \leq 270,420$
Industry maximums:	$X_4 + X_7 + X_8 \leq 405,630$ (electronics)
	$X_6 + X_9 \leq 405,630$ (oil)
	$X_5 + X_{10} \leq 405,630$ (retail sales)
Growth maximum:	$X_4 + X_5 + X_6 + X_7 \leq 540,840$
Foreign limitation:	$X_1 + X_2 + X_3 \geq 3(X_6 + X_7)$ or $X_1 + X_2 + X_3 - 3X_6 -$ $3X_7 \geq 0$
Income minimum:	$X_8 + X_9 + X_{10} \geq 540,840$
Objective function: Maximize	$.0774X_1 + .0803X_2 + .0832X_3 + .15X_4 + .18X_5 +$ $.20X_6 + .16X_7 + .12X_8 + .10X_9 + .09X_{10}$

*270,420 is 20 percent of 1,352,100.
 405,630 is 30 percent of 1,352,100.
 540,840 is 40 percent of 1.352,100.

studying Table 13-2, try to formulate an objective function and constraint equations on your own.

Satisfied that her formulation was correct, Nancy rewrote the equations to include each variable in every equation (a step that facilitated data-entry procedures in her computer program) and turned to the remote terminal in her office. Within minutes,[1] she had the printout shown in Figure 13-3.

The printout showed Nancy that in addition to the 10 variables representing stocks and bonds in the proposed portfolio, 19 additional variables had been created to facilitate the solution. The *surplus* variables, 11 and 12, were required for the greater-than constraints dealing with the offset of foreign investments by Treasury Bonds and the minimum amount to be invested in income producing stocks. The 15 *slack* variables, 13 to 27, were used in the 15 less-than constraints starting with the limitation on total funds and ending with the maximum to be invested in growth stocks. The two *artificial* variables, 28 and 29, were used with two greater-than

[1] The total time to enter data and solve this problem with a BASIC language program on a Hewlett-Packard 3000 computer was 12 minutes, only 45 seconds of which was actual computational time in the computer's central processing unit.

```
YOUR VARIABLES 1    THROUGH 10
SURPLUS VARIABLES 11   THROUGH 12
SLACK VARIABLES 13   THROUGH 27
ARTIFICIAL VARIABLES 28    THROUGH 29


ANSWERS:
  VARIABLE       VALUE
  1              120420
  14             29580
  2              90000
  6              90140
  17             90140
  8              225350
  19             180280
  20             270420
  21             45070
  23             225350
  10             45070
  26             90140
  25             45070
  5              270420
  4              180280
  3              60000
  9              270420
DUAL VARIABLES:
  COLUMN         VALUE
  11             2.66667E-02
  12             1.40667E-02
  13             .104067
  14             0
  15             .0029
  16             5.79998E-03
  17             0
  18             .06
  19             0
  20             0
  21             0
  22             9.99999E-03
  23             0
  24             .03
  25             0
  26             0
  27             1.59333E-02
OBJECTIVE FUNCTION VALUE = 173425.
```

Figure 13-3   Linear-programming computer printout.

constraints. (Artificial variables are merely computational conveniences required by the simplex algorithm and have no literal interpretation.)

After studying the printout, Nancy prepared the memo shown in Figure 13-4. Before reading this memo, examine the printout and decide how you would have responded to Bob's original note.

Bob was surprised to receive such a definitive recommendation in such a short time. He made some spot checks to satisfy himself that the conditions he had set were met and then stopped by Nancy's office to compliment her on her thorough work and to find out how she had done it. He remembered the application of linear programming to portfolio selection when she mentioned it but did not fully understand the printout she showed him. Specifically, he did not know the significance of *dual variables* in portfolio analysis. Before reading their conversation on dual variables, try to explain them to yourself.

OFFICE MEMORANDUM

TO: Bob Butcher

FROM: Nancy Simpson

SUBJECT: O'Leary Portfolio

DATE   Dec. 5, 1980

Following the guidelines in your memo of December 2, I recommend the following investments for Mr. O'Leary:

U.S. Treasury Bond, February 1993	120,420
U.S. Treasury Bond, August 1992	90,000
U.S. Treasury Bond, May 1990	60,000
Computer Technology Corporation	180,280
Goode's Discount Stores, Inc.	270,420
Persian Gulf Oil	90,140
Yakamura Electronics	None
National Transistor Corporation	225,350
Oklahoma Oil Company	270,420
Family Catalog Sales	45,070
Total	$1,352,100

If the expected annual yields do not change, this portfolio will bring an annual return of $173,425. The average annual yield will be 173,425/ 1,352,100, or approximately 12.8 percent

---

Figure 13-4   Memorandum from Nancy Simpson to Bob Butcher.

TABLE 13-3 Dual variables

Column	Constraint relaxed by $1	Increases objective function by
11	Foreign investments	.0267
12	Income stock	.0141
13	Total funds	.1041
14	February 1993 Treasury Bonds	0
15	August 1992 Treasury Bonds	.0029
16	May 1990 Treasury Bonds	.0058
17	Computer Technology Corporation	0
18	Goode's Discount Stores	.0600
19	Persian Gulf Oil	0
20	Yakamura Electronics	0
21	National Transistor Corporation	0
22	Oklahoma Oil Company	.0010
23	Family Catalog Sales	0
24	Electronic industry	.0300
25	Oil industry	0
26	Retail sales industry	0
27	Growth stock	.0159

Nancy: The dual solution to a linear-programming problem treats constraints as if they were variables. The final matrix of the dual solution includes 17 columns, numbered 11 to 27, for the 17 constraint equations (see Table 13-2). The value of the dual variable represents the amount by which the objective function would be increased if the constraint were relaxed by $1; that is, if a less-than constraint could be $1 more or a more-than constraint could be $1 less. I've jotted down the interpretation of the dual variables and rounded the values to four decimals (see Table 13-3).

Bob: I see, but some of the values are not logical. Why would we get back only .1041 on $1 of total funds when you computed the overall rate of return to be 12.8 percent?

Nancy: Remember that we are dealing with only one constraint, total funds. We have not relaxed the other constraints. The high-yield investments are already at their limits, and the extra dollar will be forced into a lower-yield investment. For example, Goode's Discount at 18 percent is at the maximum of $270,420, but Family Catalog at 9 percent is only at $45,070.

Bob: That makes sense. But why are some investments *zero?* Surely we would realize some benefit from an extra dollar in, say, Yakamura Electronics.

Nancy: Of course we would. But the dual does not represent a dollar *invested,* only a dollar *available.* Since we already have unused investment opportunity in Yakamura Electronics, relaxing the maximum investment in that firm wouldn't add anything to income.

Bob: It still seems strange that *all* these dual-variable values are less than the overall rate of return.

Nancy: Try thinking of the dual values as *marginal* rates of return as opposed to the 12.8 percent *average* rate of return. It is not unusual for all marginal rates to be less than the average. Besides, the dual values are not as isolated as they may seem. Even though we are relaxing only one constraint at a time, there are many subtle changes in the overall mix. For example, to add $1 to one firm without raising the total will require taking $1 away somewhere else. The net effect of these changes will usually be less than the simple rate of return anticipated by the original change. That is the real value of dual variables; they show net changes.

Bob: What if I relaxed *both* a firm constraint and total-fund constraint?

Nancy (laughing):   Sorry, only one to a customer! If you make two or more changes, the interpretation of dual-variable values breaks down. For that matter, even relaxing one constraint by more than $1 could be inappropriate. The dual variables merely indicate which constraints are restricting an increase to return on investment. If we actually relax some constraints, we would have to return the problem with the revised constraints to determine the new return on investment.

Bob: By the way, how did you know that the dual variable in column 11 refers to the income-stock minimum, and so on?

Nancy: Having used this program before, I know how it works, but there are also clues to the identity of dual variables in the designation of variables at the top of the printout. Since surplus variables are used only with greater-than constraints, I know that variables 11 and 12 and therefore columns 11 and 12 refer to the constraints on foreign investment and income stock, respectively, because that is the order in which I entered them. Similarly, slack variables are used with less-than constraints, so variables (and columns) 13 to 27 refer to the 15 less-than constraints, again in the order in which they were entered (see Table 13-2). The artificial variables, 28 and 29, are used in the more-than constraints. However, since the more-than constraints are already represented by the surplus variables, there is no need for a 28th or 29th column.

Bob: Amazing! Nancy, you're too smart for me. I'm glad that this trust is blind and O'Leary doesn't know what a good job you've done, otherwise he might want to take you to Washington with him to straighten out some of the problems there!

CHAPTER 14

Marketing Management Applications

THE LEMON DELIGHT CASE

About 15 months ago, Amalgamated Foods, Inc., introduced what proved to be a highly successful powdered lemonade mix called Lemon Delight. Encouraged by grocery store sales, Amalgamated is attempting to perfect a ready-to-drink version of Lemon Delight to be sold in standard 12-ounce soft drink cans. Cans will make Lemon Delight available in vending machines, convenience stores, amusement parks, and other markets not accessible through the powdered version.

Although there is a very small amount of natural lemon product in Lemon Delight, the presence of citric acids requires certain preservatives to be added in the ready-to-drink version. Anxious to preserve the natural flavor that made powdered Lemon Delight a success, and in spite of the chemists' insistance that the additives are undetectable, Amalgamated has decided to make a comparative taste test of the two drinks.

Six hundred and twenty-five students from a nearby community college have volunteered to participate in the test. Since powdered Lemon Delight is available in dispensers in the college dining hall, all the students in the test feel that they are familiar with its taste. The test consists of giving each student two coded cups, one containing 3 ounces of chilled Lemon Delight made from the powdered mix and the other containing the same amount of similarly chilled canned Lemon Delight. The students are asked to taste each and to identify the original, powdered version of Lemon Delight.

Before considering the results, let us examine the logic used by the quality control department in designing the test. Essentially, this is a test of a hypothesis about the population of Lemon Delight drinkers. As in previous hypothesis testing, the appropriate null hypothesis is that there is no change in the population mean. But what do sample data suggest about such a population? A sample of weights or lengths gives an estimate of the population weight or length, but what does it mean if, say, 300 of 625 students identify the powdered drink mix? Think about this for a moment before reading on.

An important assumption in this problem is that guessing will result in as many right answers as wrong answers (something like taking a true-false test on a totally unknown subject). Thus, if there is no taste difference between canned and powdered Lemon Delight, we would expect 50 percent to guess correctly and 50 percent to guess incorrectly in an identification test. Notice that the number of guesses is expressed as a *percentage;* that is the only way we can relate sample results to population data in this problem. A percentage, of course, is the same as a *proportion* and we recall the discussion of proportions at the end of Chapter 4.

We can now state a null hypothesis about the mean proportion of a population:

$$H_0: \quad \pi = .50$$

It is standard practice at Amalgamated to test for quality control at the .05 level of significance. We therefore add two additional items of information necessary for a hypothesis test

$$\alpha = .05$$

and because the distribution of proportions is a *normal approximation* of a binomial distribution and the statement of the null hypothesis as an equality suggests a *two-tail test,* we can also state that

$$z = \pm 1.96$$

Although we know that proportions are normal approximations of binomial distributions, we should confirm that all normal and binomial conditions are met in this particular example:

- The guess is either right or wrong, the test can be conducted so that guesses are independent (students are not aware of the guesses of others), and over the short period of the test there is no reason to believe that the probability of a correct guess will change. We therefore have a *Bernoulli* process. The distinct number of trials further makes this a *binomial* process.

- The sample size (625) is large; that is, $n \geq 30$.

- Both *np and n(1 − p) exceed 5*. Actually, in this case, we use the hypothesized value of π for p and compute $n\pi$ as $(625)(.5) = 312.5$ and $n(1 − \pi)$ as $(625)(1 − .5) = 312.5$.

Since the distribution of proportions will be approximated with the normal, we shall need the equivalents of normal distribution parameters, the mean and standard deviation or error:

$$\text{Mean proportion} = \pi = .50 \qquad (4\text{-}8)$$

$$\text{Standard error of proportion} = \sigma_\pi = \sqrt{\frac{\pi(1 − \pi)}{n}} \qquad (4\text{-}9)$$

$$= \sqrt{\frac{(.5)(1 − .5)}{625}}$$

$$= \sqrt{.0004} = .02$$

Now the problem becomes a fairly routine application of hypothesis testing with a normal distribution. The hypothesis has already been formulated and the significance level established. The next step is to formulate a decision rule.

The generalized decision rule in this case is

Accept the null hypothesis if the sample proportion is within 1.96 standard errors of the proportion from the mean.

For the specific values of the mean and standard error of the proportion in this case, the decision rule is

Accept if the sample proportion (p) is greater than the hypothesized mean (.50) less 1.96 times .02 or less than .50 plus 1.96 times .02.

That is, the critical values are

$$CV = \pi \pm z\sigma_\pi$$
$$= .50 \pm (1.96)(.02)$$
$$= .50 \pm .0392$$
$$= .4608 \text{ and } .5392$$

If the actual test results in 300 correct guesses out of 625, what should Amalgamated conclude? What if 350 students correctly identify the canned drink? Try to answer these questions before consulting Figure 14-1.

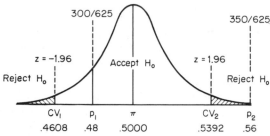

Figure 14-1 Test of the null hypothesis that the population proportion π is equal to .5. The sample proportion p_1 of 300/625 = .48 falls in the acceptance region; therefore, H_0 cannot be rejected. The conclusion is that there is no discernable difference between the taste of powdered Lemon Delight and the taste of the canned variety. The sample proportion p_2 of 350/625 = .56 falls in the rejection region; therefore, H_0 must be rejected. The conclusion is that the canned and powdered drinks have significantly different tastes at the .05 level.

The actual test resulted in 324 correct guesses and 301 incorrect guesses. The sample proportion of 324/625 = .5184 fell well within the critical values of the hypothesis test. Amalgamated correctly concluded that the additives did not alter the taste of Lemon Delight, the chemists somewhat smugly said, "We told you so," and the canned version of Lemon Delight was marketed with a success nearly equaling that of the powdered mix.

THE PASSION PULP PUBLISHERS CASE

Jack Henderson sat musing in his office. His recent appointment as head of one of the paperback divisions of a large New York publishing house was clearly a promotion, but it was hardly a direct route to the top executive position he had set as his goal. Not that being a division head was all that bad, it was *this* division, jokingly referred to by his colleagues as Passion Pulp Publishers, or simply PPP. It would take a truly spectacular performance in PPP to bring himself to the attention of top management. Grimly, he reviewed the situation.

PPP specialized in paperback romance novelettes. Manuscripts were received from free-lance writers, English faculty from nearby universities (invariably under pen names), and from contract authors who reworked older stories into contemporary settings. A reviewer screened manuscripts and passed on those considered worthy, giving each acceptable manuscript a rating between 1 and 10 to reflect its relative merit.

Cover design and artwork were done by the two company artists who, although they were full-time employees, spent only about 10 percent of their time on PPP jobs. Two writers, who were also full-time with the company but spent only a few hours per week on PPP work, prepared brief synopses to be printed on the back cover of each paperback.

The novelettes were distributed to participating supermarket chains, which usually displayed them on revolving wire racks near the checkout counters. The idea was to attract customers in the checkout line with an intriguing cover and to sell them with an interest-arousing synopsis. The distribution agreement permitted grocers to return unsold novelettes for a credit after 90 days.

Jack's immediate problem lay in this return clause of the agreement. PPP was currently printing 30,000 of each novelette and often accepting 10,000 to 20,000 back. Jack looked at a summary of the last 20 novelettes and tried to think of a way to reduce the return rate. The summary is shown in Table 14-1.

Jack's first observation was that 30,000 was well over the highest sales of 23,800 for N018-78. He realized that thousands of stores contributed

TABLE 14-1   Novelette sales summary

Book number	Artist	Reviewer's rating	Synopsis writer	Net sales,* thousands
N346-77	Brian	5	Carlos	16.2
N489-77	Brian	6	Mitzi	9.0
N619-77	Sherry	5	Mitzi	14.2
N018-78	Sherry	4	Carlos	23.8
N086-78	Brian	3	Carlos	14.1
N124-78	Sherry	5	Mitzi	16.2
N199-78	Sherry	4	Carlos	19.2
N256-78	Brian	3	Carlos	17.4
N307-78	Sherry	6	Carlos	19.6
N339-78	Brian	1	Mitzi	7.1
N421-78	Sherry	4	Carlos	22.7
N504-78	Brian	1	Mitzi	11.9
N580-78	Sherry	5	Mitzi	17.9
N626-78	Brian	4	Mitzi	5.4
N684-78	Brian	7	Carlos	15.9
N736-78	Sherry	5	Mitzi	16.7
N075-79	Brian	6	Mitzi	12.2
N115-78	Brian	3	Carlos	14.6
N174-79	Sherry	7	Carlos	21.5
N223-79	Sherry	4	Mitzi	15.4

*Net sales = total printing − returns.

to those sales and that some of those stores had probably sold all of their copies of N018-78, even though over 6000 had been returned by other stores. Nonetheless, he was willing to sacrifice some sales for reduced returns. He finally decided that he would accept a 10 percent chance that demand would exceed supply. He still planned to keep volume the same for all books, but what volume would satisfy his 10 percent stock-out criterion? Before reading on, ask yourself how you would solve this problem using statistical concepts.

Jack realized that sales demand followed a probability distribution, but only if it followed a known theoretical distribution could he solve the problem by the technique he had in mind. Feeling that a graphical presentation would be easier to interpret than the summary table he was using, he made a histogram of sales, as shown in Figure 14-2.

Jack's histogram showed the greatest frequency of sales in the center class and smaller frequencies toward the extremes. He knew that a sample of only 20 would not produce a smooth curve, but he felt that the general shape of his histogram justified the use of the normal distribution.[1]

Using a hand-held calculator that was preprogrammed to give simple statistical measures, he found that the mean of his sample of sales figures was 15.55, or 15,550, and the sample standard deviation was 4.82, or 4820. If you do not have such a calculator, you can confirm these values as follows:

$$\text{Sample mean} = \overline{X} = \frac{\Sigma X}{n} \tag{5-1}$$

$$= \frac{16.2 + 9.0 + 14.2 + \cdots + 15.4}{20}$$

$$= \frac{311}{20} = 15.55$$

$$\begin{array}{l}\text{Sample} \\ \text{standard deviation} =s = \sqrt{\dfrac{\Sigma\,(X - \overline{X})^2}{n - 1}} \end{array} \tag{5-2}$$

$$= \sqrt{\frac{(16.2 - 15.55)^2 + (9.0 - 15.55)^2 + \cdots + (15.4 - 15.55)^2}{20 - 1}}$$

$$= \sqrt{\frac{(.65)^2 + (-6.55)^2 + \cdots + (-.15)^2}{19}}$$

$$= \sqrt{\frac{.4225 + 42.9025 + \cdots + .0225}{19}}$$

$$= \sqrt{\frac{442.27}{19}} = \sqrt{23.2774} = 4.82$$

[1]There are standard statistical methods, such as the *chi-square goodness-of-fit test*, for testing the applicability of a theoretical distribution to sample data. An intermediate college-level statistics book may be consulted for the computational procedures required for such a test.

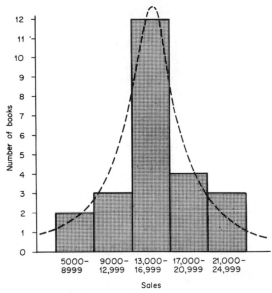

Figure 14-2 Histogram of paperback novelette sales.

Jack sketched a normal distribution with these parameters and studied it. He visualized the supply of novelettes as a vertical line on his sketch. As he moved the line to the right, increasing the supply, the probability that demand would exceed supply, represented by the area under the curve to the right of the supply line, decreased. Jack's sketch is shown in Figure 14-3.

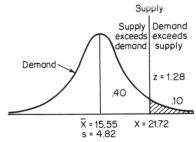

Figure 14-3 A 10 percent probability that demand exceeds supply.

Having set the risk that demand would exceed supply at 10 percent, Jack computed the value of supply that would satisfy this condition. The area between the mean and the supply level must equal $.50 - .10 = .40$. In a normal distribution, this corresponds to a number of standard deviations, called z, of 1.28:

$$z = \frac{X - \overline{X}}{s} \tag{4-7}$$

$$1.28 = \frac{X - 15.55}{4.82}$$

$$X = (1.28)(4.82) + 15.55$$
$$= 6.17 + 15.55$$
$$= 21.72 \text{ or } 21,720 \text{ copies of each novelette[2]}$$

Jack felt rather pleased with himself for his scientific approach to this problem but decided that he would seek his assistant's opinion of his solution. He knew that his assistant, Willy Franklin, had a stronger background in statistics than he had. If Willy had a better solution, he would certainly apply it; if not, he just might gain Willy's respect for his own statistical prowess.

Willy politely agreed with Jack's use of the normal distribution to determine the number of copies to be printed but pointed out that there appeared to be factors other than random chance contributing to demand level. If these factors were considered, perhaps a better estimate of demand for each novelette could reduce returns even further.

"I suppose you are thinking of a regression analysis," said Jack. "I considered that, but I don't see how we could use any factor other than the reviewer's rating to predict sales, and I have a feeling that those ratings don't have much to do with sales."

"They may or may not," said Willy, "the regression analysis will tell us that. What I have in mind is the relationship between sales and who does the cover and synopsis."

Jack agreed that those were probably the major determinants of sales but failed to see how people could be incorporated into a quantitative analysis.

"Ordinarily, they can't," said Willy, "but in this case, because you have only two persons in each job, they can. Watch."

What Willy did is shown in Table 14-2.

"I see that you have assigned Brian and Carlos the number 2 and Sherry and Mitzi the number 1," said Jack. "Are you telling me *that* is the way to quantify data? It seems a little simplistic to me."

"It *is* simplistic," said Willy, "but it works."

"Does this mean that you think that Brian and Carlos are twice as good as Sherry and Mitzi?"

"No. The numbers are arbitrary. As long as they are different, the

---

[2]Formula (4-7) actually reads $z = (x - \mu)/\sigma$. The formula given here is an alternate form in which the sample statistics $\hat{X}$ and s are substituted for the population-parameter counterparts μ and σ, respectively.

TABLE 14-2 Willy's revised sales summary

Book number	Artist	Reviewer's rating	Synopsis writer	Net sales,* thousands
N346-77	Brian 2	5	Carlos 2	16.2
N489-77	Brian 2	6	Mitzi 1	9.0
N619-77	Sherry 1	5	Mitzi 1	14.2
N018-78	Sherry 1	4	Carlos 2	23.8
N086-78	Brian 2	3	Carlos 2	14.1
N124-78	Sherry 1	5	Mitzi 1	16.2
N199-78	Sherry 1	4	Carlos 2	19.2
N256-78	Brian 2	3	Carlos 2	17.4
N307-78	Sherry 1	6	Carlos 2	19.6
N339-78	Brian 2	1	Mitzi 1	7.1
N421-78	Sherry 1	4	Carlos 2	22.7
N504-78	Brian 2	1	Mitzi 1	11.9
N580-78	Sherry 1	5	Mitzi 1	17.9
N626-78	Brian 2	4	Mitzi 1	5.4
N684-78	Brian 2	7	Carlos 2	15.9
N736-78	Sherry 1	5	Mitzi 1	16.7
N075-79	Brian 2	6	Mitzi 1	12.2
N115-78	Brian 2	3	Carlos 2	14.6
N174-79	Sherry 1	7	Carlos 2	21.5
N223-79	Sherry 1	4	Mitzi 1	15.4

*Net sales = total printing − returns.

regression-equation coefficients will tell us which ones are better. However, if we had *three* artists, the middle number would imply a center position. This technique will work when a variable can take on only two values, such as male or female, college graduate or nongraduate, or, as in this case, Brian or Sherry and Carlos or Mitzi."

"OK, if you say so. But I'd like to see the results of such an analysis before we make any decisions."

"I'll have it for you in an hour."

Even Willy made mistakes occasionally, and it was closer to 2 hours later that he brought in the printout shown in Figure 14-4.

"If I recall correctly the meaning of these things, the R-SQUARED is telling me that we can account for almost 85 percent of the variation in sales by the reviewer's rating plus the persons who do the cover and synopsis," said Jack. "Is that right?"

"That's right," said Willy. "Here's the regression equation"[3]

$$Y = a + b_1X_1 + b_2X_2 + b_3X_3 \tag{7-11}$$
$$= 15.7588 - 6.25485X_1 + .0851136X_2 + 5.86599X_3$$

[3]With the addition of b_3X_3 for the third independent variable.

```
INDEX              MEANS              STANDARD DEVIATIONS
  1                1.5                 .512989
  2                4.4                1.66702
  3                1.5                 .512989
  4               15.55               4.82467


INDEX              B                  STD. ERROR       T-RATIO
  0               15.7588             2.54389          6.19476
  1               -6.25485             .961897        -6.50262
  2                8.51136E-02         .298271          .285356
  3                5.86599             .922233         6.36064


R-SQUARED= .84873                     R= .921265


CORRELATION COEFFICIENTS
  1                -.307729            0                -.674104

 -.307729          1.                  .123092           .310838

  0                 .123092            1                 .627329

 -.674104           .310838            .627329          1.


ACTUAL             PREDICTED          RESIDUAL
 16.2              15.4066             .793362
  9                 9.62576           -.625762
 14.2              15.7955           -1.5955
 23.8              21.5764            2.22362
 14.1              15.2364           -1.13641
 16.2              15.7955            .4045
 19.2              21.5764           -2.37638
 17.4              15.2364            2.16359
 19.6              21.7466           -2.14661
  7.1               9.2002           -2.10019
 22.7              21.5764            1.12362
 11.9               9.2002            2.69981
 17.9              15.7955            2.1045
  5.4               9.45554          -4.05554
 15.9              15.5769            .323133
 16.7              15.7955            .9045
 12.2               9.62576           2.57424
 14.6              15.2364           -.636412
 21.5              21.8317           -.331718
 15.4              15.7104           -.310388


STAND. ERROR OF EST.= 2.04485                    D.F.= 16
```

Figure 14-4 Multiple-regression computer printout.

"For example," Willy continued, "if Brian does the cover, the reviewer rates the book 6, and Mitzi writes the synopsis, as happened on book N489-77, you can expect sales of 9.62576 thousand or about 9626."

Mostly to kill time while he considered this, Jack checked this figure on his calculator after rounding off the regression coefficients:

$$Y = 15.76 - 6.25X_1 + .09X_2 + 5.87X_3$$
$$= 15.76 - (6.25)(2) + (.09)(6) + (5.87)(1)$$
$$= 15.76 - 12.50 + .54 + 5.87$$
$$= 9.67 = 9670$$

"Yes, that's about what I get," said Jack. "Is this the amount we should print for this combination?"

"Well, you can see that the actual sales were only 9000 and you would have printed about 670 too many with that number. But that's much better than the 21,000 excess we printed under the current system and the 12,720 excess that would have resulted had we used your univariate estimate. However, you had a good idea in incorporating a 10 percent chance of demand exceeding supply. We can still do that, but now we'll use the standard error of the estimate instead of the standard deviation and your predicted value of sales instead of the mean:"[4]

$$t_{(\alpha,\, df)} = \frac{Y - \hat{Y}}{S_{Y|X_1X_2X_3}} \tag{4-7}$$

$$t_{(.10,16)} = 1.34 = \frac{Y - 9.67}{2.04}$$

$$Y = (1.34)(2.04) + 9.67$$
$$= 2.73 + 9.67$$
$$= 12.40 \text{ or } 12,400 \text{ books}$$

"I think you may have something here, Willy, but there are a few things I still don't understand. First the Y intercept is supposed to represent zero values of the independent variables. How can we have zero for the cover and synopsis? Second, the negative coefficient of X_1 implies that the covers are *reducing* our sales; I refuse to believe this!"

Can you answer these questions? Here's how Willy did.

"Good questions, and they are related. Of course, the Y intercept should not be interpreted in terms of no cover and no synopsis. And covers don't decrease sales. The negative coefficient of X_1 simply means that

---

[4]This is still another variation of the basic formula, $z = (X - \mu)/\sigma$. This time, in addition to the substitutions for μ and σ already noted, the t distribution is used, and the random variable is designated Y instead of X.

sales are inversely related to the arbitrary values assigned to Brian and Sherry. In other words the higher value, 2 for Brian, results in fewer sales than the lower value, 1 for Sherry. Review ratings and synopsis are *directly* related; the higher the value, the greater the sales. The Y intercept simply is the computational result of fitting these numbers, along with actual sales, to a mathematical equation. If you must interpret it, you can think of it as being necessary to offset the negative coefficient of X_1. Incidently, these relationships are also borne out in the simple coefficients of correlation. X_1 to Y is negative, $-.674104$; X_2 to Y and X_3 to Y are positive, $.310838$ and $.627329$, respectively."

"Can I conclude that Sherry does better covers than Brian and that Carlos writes better synopses than Mitzi?"

"Let's see . . . for covers, the inverse relationship means that 1, Sherry, results in higher sales and for synopses, the direct relationship means that 2, Carlos, results in higher sales . . . yes, that's right."

"Hmm. That gives me something else to think about. Anything more I should know?"

"Well, yes. Look at the t ratios. Everything but the coefficient of X_2 appears to be significant. The t ratio of only $.285356$ means that the coefficient of $.0851136$ is not significantly different from zero. This confirms your suspicion that the review ratings are not all that helpful in predicting sales. The low simple correlation coefficient suggests this, too."

"What is the dividing line between a significant and insignificant t ratio?"

"Oh, for 16 degrees of freedom and an alpha of .05, I'd guess about 2.00."[5]

"Good, Thanks much, Willy. I think I know exactly what to do now."

"You're quite welcome, boss. Any time."

As you may have guessed. Jack took Brian and Mitzi off PPP work, used Sherry and Carlos exclusively, fired the reviewer, and began printing paperbacks according to a 10 percent chance of a multiple-regression-predicted demand exceeding supply. Sales soared, returns dwindled, and Jack was duly rewarded with a vice-presidency in the parent firm. Willy received no credit whatsoever and is now teaching in a small college.

---

[5]Willy is a pretty good guesser. From Appendix G, the actual t value is 2.12 when α is split with .025 in each tail.

CHAPTER 15

Personnel Management Application

THE REGIONAL STATE
UNIVERSITY CASE

The director of placement services at Regional State University (RSU) has a number of responsibilities. In addition to counseling students on résumé preparation and scheduling and coordinating interviews with prospective employers, the director must keep abreast of the current starting salaries and employment prospects in the various fields of study. It is this latter responsibility that has been given increased emphasis by the current administration at RSU. Now more than ever before, prospective students (and their parents) want information on job opportunities in order to evaluate the many programs of study offered by the university. The academic programs at RSU are organized into three separate colleges, the College of Arts and Sciences, the College of Business, and the College of Engineering.

The director subscribes to a number of periodicals that report nationwide statistics concerning the employment and starting salaries of recent college graduates. In fact, the director recently received an issue of one periodical that lists the nationwide average starting salary for last year's graduates in arts and sciences, business, engineering, and several other fields. Table 15-1 summarizes the national data.

The director is hesitant to use the national data because they include graduates of many institutions that are several times larger (and more prestigious) than RSU as well as graduates of many private colleges that are very small and limited in their program offerings. As a result, it has been decided that a survey of last year's graduating class at RSU should

287

TABLE 15-1   Nationwide average starting salaries for last year's graduating class

Field of study	Nationwide average starting salary
Arts and sciences	$11,100
Business	$11,800
Engineering	$12,300

be conducted. The director feels that the survey will provide starting-salary data relevant to students at RSU and also show whether RSU graduates receive starting salaries in line with the national averages.

In planning the study, the director jotted down the following list of questions that should be answered from the data:

1. What is the *overall* average starting salary for last year's graduating class at RSU?

2. What is the average starting salary for *women* in last year's graduating class?

3. What is the average starting salary for *men* in last year's graduating class?

4. Did last year's arts and sciences graduates of RSU start at the national average ($11,100)?

5. Did last year's business graduates of RSU start at the national average ($11,800)?

6. Did last year's engineering graduates of RSU start at the national average ($12,300)?

The director felt strongly that telephone interviews with the graduates were necessary in order to obtain the salary information needed. It was felt that mail questionnaires would simply be too impersonal and would not yield a very good response rate. In a telephone interview, the graduate could be told why the information was needed and could be assured in much more personal terms that it would be kept confidential.

The director was feeling very good about the project until the computer printout listing the individuals in last year's graduating class arrived from Alumni Services. The printout listed exactly 1108 names. It suddenly became obvious that the time and expense of calling all those people was well beyond the resources available for the project. Sitting alone thumbing through the printout, the director was privately embarrassed about failing to consider the large number of graduates.

It wasn't long, however, before an assistant came into the office and suggested to the director that it was not necessary to contact *all* the graduates in order to obtain reliable data on starting salaries. In fact, all the questions on the director's list could be answered by interviewing a *sample* of the graduates. The assistant pointed out that sampling might lead to errors but that the magnitude of the errors could be estimated from the data. The assistant went on to remind the director that the nationwide data summarized in Table 15-1 were most certainly collected through some type of sampling procedure.

In a matter of seconds the assistant was given responsibility for conducting the survey. Before reading how the data were collected and analyzed, look back over the list of questions prepared by the director and try to identify the statistical concepts and techniques that will be needed.

The Sample

Given the financial resources and time limitations of the project, the assistant decided to interview a sample of 36 members of last year's graduating class. The sample graduates were randomly selected using a table of random digits. Since each student at RSU is given a unique four-digit identification number, the assistant was able to select four-digit numbers from a random-number table (such as the one in Appendix H) and use each to identify a student in last year's graduating class. In the event that a random number was generated that did not coincide with an ID number of a graduate, the number was simply discarded and the next random number was generated. The random selection procedure ensured that each of the 1108 graduates had an equal chance of being included in the sample to be interviewed.

The sample-selection procedure went smoothly, and within 3 weeks the 36 graduates in the sample had been reached and interviewed. For each graduate in the sample, information was recorded concerning academic field, sex, and starting salary. The sample data are recorded in Table 15-2. While looking over the data, the assistant referred back to the original list of questions the director had compiled for the study. We shall now analyze the data and answer those questions.

The Statistical Analysis

Question 1: What is the overall average starting salary for last year's graduating class at RSU?

TABLE 15-2   The sample data

Student ID number	Academic area	Sex	Starting salary
1881	Botany (A&S)	M	$ 9,750
3214	Economics (B)	M	10,150
2186	Industrial engineering (E)	M	13,250
4211	Math (A&S)	F	11,275
1211	General business (B)	M	11,250
9210	Journalism (A&S)	F	9,950
8218	Psychology (A&S)	M	10,150
7235	General business (B)	M	10,250
6918	Physics (A&S)	M	13,100
8335	Accounting (B)	F	13,125
9654	Spanish (A&S)	M	8,750
2231	Mechnical engineering (E)	M	12,950
7652	Biology (A&S)	F	10,000
8131	Chemistry (A&S)	M	12,500
6621	General business (B)	F	10,200
2351	Accounting (B)	M	14,800
5232	History (A&S)	F	10,150
5129	Industrial engineering (E)	M	11,950
2159	Civil engineering (E)	M	12,350
6223	Psychology (A&S)	F	12,900
8354	General business (B)	F	11,200
1962	Civil engineering (E)	M	11,900
2851	Journalism (A&S)	F	11,000
1624	Industrial engineering (E)	M	11,900
8111	Accounting (B)	F	12,750
9241	Industrial engineering (E)	M	13,200
4434	Biology (A&S)	M	10,250
2734	English (A&S)	F	10,125
1814	Accounting (B)	M	11,975
2214	Industrial engineering (E)	M	12,950
8814	Chemistry (A&S)	F	11,500
7676	Math (A&S)	F	10,600
8291	General business (B)	F	11,200
1400	English (A&S)	M	10,900
2229	Civil engineering (E)	M	12,900
1965	Economics(B)	F	11,100

The sample data are summarized by computing the mean and standard deviation of the 36 graduates who were interviewed:

$$n = 36$$

$$\overline{X} = \frac{\Sigma X}{n} = \$11,506.94 \tag{5-1}$$

$$s = \sqrt{\frac{\Sigma(X - \overline{X})^2}{n - 1}} = \$1347.20 \tag{5-2) or (5-3}$$

Thus, the best single value (point estimate) that can be used to estimate the average starting salary for last year's graduating class is $11,506.94.

If all 1108 graduates had been interviewed, the computation of the average starting salary would answer the question completely. However, since only 36 graduates provided information for the study, it is quite possible that the population mean, that is, the mean salary for the entire graduating class, is different from $11,506.94. It would be helpful, then, to develop an interval estimate that shows a range of salaries within which we are confident that the true average of last year's graduating class will fall.

Let us develop an interval estimate at the .95 confidence level. A confidence level of .95 means that there is a .95 probability that the true average starting salary lies within the lower and upper limits of the interval. Since we are working with a large sample ($n \geqslant 30$) and a population whose standard deviation is unknown, the interval estimate is given by

$$\overline{X} \pm z s_{\overline{X}} \qquad (6\text{-}2)$$

where $\overline{X}$ = the sample mean

z = 1.96 standard deviations (Appendix F)

$s_{\overline{X}}$ = estimate of standard error of mean

To compute the interval, the standard error of the mean is estimated as

$$s_{\overline{X}} = \frac{s}{\sqrt{n}} = \frac{\$1347.20}{\sqrt{36}} = \$224.53$$

and the interval estimate at the .95 confidence level is

$$\$11,506.94 \pm 1.96\ (\$224.53)$$
$$\$11,506.94 \pm \$440.09$$
$$\$11,066.85 \text{ to } \$11,947.03$$

Thus, there is a .95 probability that the average starting salary for last year's graduating class at RSU is between $11,066.85 and $11,947.03. As an alternative explanation, it can be said that of all the possible samples of 36 students that could be drawn from the graduating class of 1108, 95 percent of the sample average salaries would fall between $11,066.85 and $11,947.03.

Question 2: What is the average starting salary for females in last year's graduating class?

Table 15-2 shows that 15 of the 36 graduates in the sample are females. The data for these 15 female graduates are summarized below:

$$n = 15$$
$$\bar{X} = \$11,138.33 \qquad (5\text{-}1)$$
$$s = \$\ 1059.22 \qquad (5\text{-}2) \text{ or } (5\text{-}3)$$

The point estimate, then, is that female graduates of last year started at $11,138.33. Let us also develop an interval estimate, this time at the .99 confidence level. That is, we want to establish a salary range within which there is a .99 probability that the true average starting salary for females will fall. In this case the sample size is considered small ($n < 30$) since only 15 of the 36 graduates were females and the population standard deviation is again unknown. Assuming that the starting salaries for *all* female graduates at RSU last year follow a normal distribution, the interval estimate is given by

$$\bar{X} \pm ts_{\bar{X}} \qquad (6\text{-}3)$$

where t represents the number of standard errors in the t distribution (from Appendix G).

In this case,

$$t_{(.005,14)} = 2.977$$

and
$$s_{\bar{X}} = \frac{\$1,059.22}{\sqrt{15}} = \$273.49$$

The interval estimate is

$$\$11,138.33 \pm 2.977 \ (\$273.49)$$
$$\$11,138.33 \pm \$814.18$$
$$\$10,324.15 \text{ to } \$11,952.51$$

Question 3: What is the average starting salary for males in last year's graduating class?

Of the 36 sample graduates 21 are males. The data for this subsample are summarized as follows:

$$n = 21$$
$$\overline{X} = \$11,770.24 \qquad\qquad (5\text{-}1)$$
$$s = \$\ 1488.66 \qquad\qquad (5\text{-}2)\ \text{or}\ (5\text{-}3)$$

Let us again develop a .99 confidence interval around the point estimate of $11,770.24. Once again the sample is considered small, and the population standard deviation is not known. The interval estimate is again given by

$$\overline{X} \pm ts_{\overline{x}} \qquad\qquad (6\text{-}3)$$

For the male graduates, from Appendix G,

$$t_{(.005,20)} = 2.845$$

and
$$s_{\overline{X}} = \frac{\$1488.66}{\sqrt{21}} = \$324.85$$

The interval estimate at the .99 confidence level for the starting salary of male graduates is

$$\$11,770.24 \pm 2.845\ (\$324.85)$$
$$\$11,770.24 \pm \$924.20$$
$$\$10,846.04\ \text{to}\ \$12,694.44$$

Question 4: Did last year's arts and sciences graduates at RSU start at the national average ($11,100)?

Table 15-2 shows that 16 of the 36 graduates in the sample held degrees from the College of Arts and Sciences. The summary data for the starting salaries of those 16 graduates are

$$n = 16$$
$$\overline{X} = \$10,806.25 \qquad\qquad (5\text{-}1)$$
$$s = \$\ 1201.82 \qquad\qquad (5\text{-}2)\ \text{or}\ (5\text{-}3)$$

The average starting salary of $10,806.25 for the arts and sciences graduates in the sample falls below the reported national average of $11,100. However, the difference could be due to sampling error. To find out if there is a "true" difference we must conduct a hypothesis test. The procedure we shall follow for conducting the hypothesis test is described in detail in Chapter 6.

1. *Formulate the hypothesis.* The average starting salary for arts and sciences graduates last year at RSU is equal to the national average of $11,100

$$H_0: \mu = \$11,100$$

2. *Establish a significance level.* Suppose that we want only a .05 probability of rejecting H_0 (concluding that RSU salaries for arts and sciences graduates differ from the national average) when it is true (the salaries do not differ from the national average)

$$\alpha = .05$$

3. *Formulate a decision rule.* This test is two-tailed; the average salary of RSU graduates in arts and sciences could be above or below the national average. The hypothesis should be accepted if the sample average falls within an interval estimate around the national average at the $1 - \alpha$ (.95 in this case) confidence level. The sample is small ($n <$ 30), and the population standard deviation is unknown. Assuming that the starting salaries of all arts and sciences graduates in last year's class follow a normal distribution, the critical values in the interval estimate are, from Table 6-3,

$$CV = \mu \pm ts_{\bar{X}}$$

where μ is the reported national average of $11,100.

To compute the interval, from Appendix G,

$$t_{(.025,15)} = 2.131$$

$$s_{\bar{X}} = \frac{\$1201.82}{\sqrt{16}} = \$300.46$$

and

$$CV = \$11,100 \pm 2.131(\$300.46)$$
$$= \$11,100 \pm \$640.28$$
$$= \$10,459.72 \text{ to } \$11,740.28$$

The decision rule is

If the average starting salary of the 16 arts and sciences graduates in the sample falls between $10,459.72 and $11,740.28, accept the hypothesis that RSU arts and sciences graduates start at the national average. If not, reject the hypothesis.

4. *Take a sample.* The sample data show an average starting salary of $10,806.25 for the 16 arts and sciences graduates.

5. *Make a decision.* Since the sample average of $10,806.25 falls within the critical values of $10,459.72 and $11,740.28, accept the hypothesis. Arts and sciences graduates in last year's class at RSU received starting salaries in line with the national average.

Question 5:   Did last year's business graduates at RSU start at the national average ($11,800)?

Table 15-2 shows 11 business graduates in the sample of 36. Once again, the statistical calculations will be based on a small sample. The starting salaries for the 11 business graduates are summarized below:

$$n = 11$$
$$\bar{X} = \$11,636.36 \tag{5-1}$$
$$s = \$1439.99 \tag{5-2} \text{ or } (5-3)$$

The hypothesis test is necessary to determine whether the sample average of $11,636.36 is "significantly different" from the national average of $11,800.

1. *Formulate the hypothesis*

$$H_0: \mu = \$11,800$$

2. *Establish a significance level*

$$\alpha = .05$$

3. *Formulate a decision rule.* The critical values are determined by a .95 interval estimate around the national average of $12,500. The interval is given by (Table 6-3)

$$CV = \mu \pm ts_{\bar{X}}$$

In this case, from Appendix G,

$$t_{(.025,10)} = 2.228$$

$$s_{\bar{X}} = \frac{\$1434.99}{\sqrt{11}} = \$432.67$$

and
$$CV = \$11,800 \pm 2.228(\$432.67)$$
$$= \$11,800 \pm \$963.99$$
$$= \$10,836.01 \text{ to } \$12,763.99$$

The decision rule is

> *Accept the hypothesis if the sample average falls between $10,836.01 and $12,763.99. Otherwise, conclude that the business salaries for last year's RSU graduates are not in line with the national average.*

4. *Take a sample*

$$\overline{X} = \$11,636.36$$

5. *Make a decision.* Since $\$10,836.01 < \$11,636.36 < \$12,763.99$, accept H_0. Last year's business graduates also received starting salaries in line with the national average.

Question 6:   Did last year's engineering graduates at RSU start at the national average ($12,300)?

Refer again to the data in Table 15-2; there are nine engineering graduates in the sample. Their starting salaries are summarized as follows:

$$n = 9$$
$$\overline{X} = \$12,594.44 \qquad\qquad (5\text{-}1)$$
$$s = \$568.14 \qquad\qquad (5\text{-}2) \text{ or } (5\text{-}3)$$

Since the sample average in this case is above the national average, let us modify the question slightly. Suppose we want to know if it would be accurate to report in a brochure describing the engineering program at RSU that "our graduates receive starting salaries higher than the national average." The hypothesis test then becomes one-tailed.

1. *Formulate the hypothesis*

$$H_0: \mu > \$12,300$$

2. *Establish a significance level*

$$\alpha = .05$$

3. *Formulate a decision rule.* The sample is small ($n < 30$) and the population (the engineering graduates in last year's class) standard deviation is unknown. The critical value for a one-tailed test (from Table 6-3) is

$$CV = \mu + ts_{\bar{x}}$$

In this case, from Appendix G,

$$t_{(.05,8)} = 1.860$$

$$s_{\bar{x}} = \frac{\$568.41}{\sqrt{9}} = \$189.38$$

and the critical value is

$$CV = \$12,300 + 1.860(\$189.38)$$
$$= \$12,300 + \$352.25$$
$$= \$12,652.25$$

The decision rule is

Accept the hypothesis if the sample average is greater than $12,652.25. Otherwise, reject.

4. *Take a sample*

$$\bar{X} = \$12,594.44$$

5. *Make a decision.* The sample average falls below the critical value of $12,652.25, and the hypothesis must be rejected. Although starting salaries for RSU engineering graduates are in line with the national average, it cannot be concluded that RSU engineering graduates receive starting salaries *above* the national average.

CHAPTER 16

Production Management Application

THE PHARMCO CASE

Pharmco began as a small-scale pharmaceutical manufacturing operation some 15 years ago. The company has grown rapidly and now sells over $1 million worth of pharmaceutical products annually. Most of its current business comes from private-label contracts with retail chain stores. Pharmco manufactures several types of nonprescription medicines that can be purchased over the counter in many discount stores and supermarkets.

The rapid growth in the private-label business has caused Pharmco to increase its production operations dramatically. This expansion has caused several problems for Jim Riley, manager of production control. Riley is convinced that the business has grown to the point where the company's manual system of inventory control is too slow and unreliable. All too often, last-minute changes in production schedules and delivery dates have to be made because the necessary materials are not in stock. Riley contends that current inventory records are inaccurate, causing the company to be overstocked in many items and critically understocked in others.

Last month, Riley submitted a written proposal to company president Clifford Johnson on the need for an automatic data processing system. In his proposal he noted that Johnson has often expressed the opinion that the company's future is in the private-label business and that all marketing forecasts indicate that private-label contracts will continue to expand. Riley carefully pointed out that some of Pharmco's current private-label clients have complained about delays in the delivery of their orders. Since any pharmaceutical manufacturer meeting Food and Drug Administration requirements can enter into the private-label business, Riley argued that Pharmco may jeopardize its present contracts and future growth if it does not, in his words, "clean up its act."

Johnson's general reaction to the proposal was favorable, but he thought the problem should be studied carefully before making a commitment to purchase a particular type of computer system. Being typically cautious, Johnson considered this a "big step" for the company and wanted to be sure that the computer system would meet both current and future needs. At the same time, he was also very sensitive to the idea that the production-scheduling and inventory problems were resulting in poor service to customers. Thus, there was a dilemma caused by the current need for some type of computer system and the desire of the company president to study the matter and ensure that the system purchased was the right one.

Shortly after submitting his proposal, Riley was requested by Johnson to develop a timetable for the switch to a computerized inventory control system. The timetable was to include an internal study of the company's needs, an analysis of available computer systems, the bidding, selection, and ordering process, and the actual installation and conversion of all inventory records to the computer. Johnson wanted the timetable not only as an estimate of how long it would take to convert to a computer system but also as a tentative schedule for carrying out the various phases of the project.

Riley realized that such a timetable would require inputs from a number of people within his own company as well as experts from the computer manufacturers. He also knew that coordination of the activities of these people would be necessary in order to identify the appropriate computer system and make it operational as soon as possible. He felt confident that he could make use of a well-known quantitative management technique in developing a meaningful timetable. Before reading how the timetable was established, what technique do you think is appropriate here?

Developing the Framework

Riley visualized the changeover to a computer-controlled inventory system in phases. Each phase would consist of some distinct activity that had to be completed. He felt that he could effectively depict how each phase was related to the overall project by using a PERT network. From the network (as you recall from Chapter 9), he could develop the timetable for the project.

The PERT network would also provide Riley with a communications device. He was well aware that many people in his own company would be affected by the new system, and he wanted them to know exactly what the company was doing. He felt that an easy-to-read network diagram would be the most effective way to show how and when the company planned to move into the computer age.

In order to develop the PERT network, Riley identified 10 distinct activities to be completed. Activity 1 consists of an internal study of the company's present and future needs in the area of automatic data processing. Activity 10, the last activity, involves the final transfer of records and last-minute adjustments before the new system is made fully operational. In between, a number of activities would have to be completed involving selection of the appropriate system, physical installation of the equipment, training of users, and so forth. Table 16-1 shows Jim Riley's list of activities.

Feeling satisfied that his list of 10 activities was complete and in sufficient detail, Riley set out to determine the sequential relationships between the activities and the time each would require. For each activity in Table 16-1, he determined which, if any, had to be finished before the project could move into the activity in question.

Following the PERT procedure discussed in Chapter 9, Riley next attempted to obtain an optimistic, most likely, and pessimistic time estimate for each individual activity. A number of persons provided input into this part of the data collection. For example, Riley talked to the accounting and finance people at Pharmco in order to estimate the time required to analyze the bids and make a final decision. Estimates for activities involving purchasing, installation, and training for the computer sys-

TABLE 16-1   Activity list

Activity Number	Description
1	Conduct internal study of the company's present and future needs
2	Examine computer systems currently available from the various manufacturers
3	Solicit bids from selected manufacturers
4	Analyze bids and make decision
5	Place order for system and await delivery
6	Conduct training sessions for personnel who will use the new system
7	Prepare the site for installation of the computer
8	Install the computer
9	Conduct "pilot run" to test the system
10	Transfer all records and operationalize the system

TABLE 16-2   Activity data

Activity	Preceding activity	Time estimates, months			Expected time [formula (9-1)]
		Optimistic	Most likely	Pessimistic	
Internal study	None	1	2	3	2
Examine systems	None	1	2	4	2.2
Solicit bids	Internal study; examine systems	1	1	1	1
Analyze bids and decide	Solicit bids	1	3	5	3
Order system	Analyze bids and decide	1	4	12	4.8
Conduct training sessions*	Analyze bids and decide	1	1	1	1
Prepare site	Analyze bids and decide	1	2	2	1.8
Install computer	Order system; prepare site	1	2	2	1.8
Conduct pilot run	Conduct training; install computer	1	1	1	1
Transfer all records and operationalize	Conduct pilot run	1	2	3	2

*The training sessions can be conducted before the computer is actually on the premises.

tem were developed after consultation with several computer manufacturers and some people Riley knew at other companies that had recently installed computer systems. Finally, the data presented in Table 16-2 were developed in order to construct the timetable. Riley computed the expected time for each activity using formula (9-1).

The data in Table 16-2 provide all the information needed to construct the PERT network and begin to establish a reasonable timetable for the project. Using the information Jim Riley has collected, you should attempt to construct the network and identify the critical path. You may want to refer back to Chapter 9 for a discussion of the concepts involved. The following sections will illustrate how Riley applied PERT analysis to his problem.

Developing the Timetable

From the activity list in Table 16-2 Riley developed the PERT chart that appears in Figure 16-1. He recorded the expected time (in months) by each activity arrow. Notice that he used dummy activities in several instances to illustrate precedence restrictions between activities. There are 14 numbered events in the network, each representing a milestone, or checkpoint, in the transition to a computerized inventory control system.

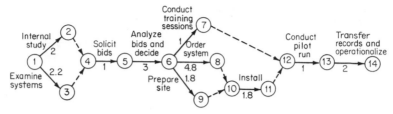

Figure 16-1   PERT network for the installation of a computer system.

Inspection of the network diagram revealed six distinct paths through the network. Each path, of course, represents a series of activities that must be completed in sequential order; a delay in any one activity would affect all that follow. Riley calculated the expected length of each path and identified the critical path as follows:

Events	Expected duration
1-3-4-5-6-8-10-11-12-13-14	15.8 months

Riley was somewhat surprised that the critical-path calculation yielded 15.8 months. When he was collecting the individual activity time estimates, he felt intuitively that the changeover could be done in about a year. However, when all the data were put together in the form of a network, it appeared that close to 16 months would be necessary. In looking back over the data, he realized that the potentially long delivery time from the computer manufacturer (the time estimates ranged from 1 to 12 months) and the time-consuming tasks of examining available computer systems and analyzing the resulting bids all served to extend the expected project length beyond what he originally thought. As he studied his PERT network more thoroughly, Jim Riley was glad that he had not made any off-the-cuff statements to the effect that the computer system could easily be in operation within a year.

In order to present Johnson with the information requested, Riley next wanted to develop a more detailed timetable that showed when the various phases of the project should be started and completed. Once management had made a commitment to switch to a computer system, this timetable would serve as a means of determining whether or not things were proceeding on schedule. Realizing that every event in the network has a specific meaning, Riley computed from his network the *earliest* time (in terms of elapsed months) each event could be expected to occur if the

TABLE 16-3   Network calculations

Event	Description	Earliest expected time T_E, elapsed months	Latest allowable time T_L, elapsed months	Slack, months
1	Beginning of project	0	0	0
2	Completion of internal study	2	2.2	.2
3	Completion of study of existing systems	2.2	2.2	0
4	Ready to solicit bids	2.2	2.2	0
5	Bids solicited	3.2	3.2	0
6	Bids analyzed and decision made	6.2	6.2	0
7	Training sessions completed	7.2	12.8	5.6
8	Computer system delivered	11	11	0
9	Site prepared for installation	8	11	3
10	Ready to install computer	11	11	0
11	Computer installed	12.8	12.8	0
12	Ready to begin pilot run	12.8	12.8	0
13	Pilot run completed	13.8	13.8	0
14	Inventory records transferred and system made operational	15.8	15.8	0

project remained on schedule. He then computed the *latest* time each event could occur (again expressed in elapsed months) if the new system was to be in operation after 15.8 months. His calculations are summarized in Table 16-3. You should verify them.

Riley's computations in Table 16-3 show that three phases of the project, the internal study, the training sessions, and the preparation of the site, could extend varying degrees beyond their expected completion without slowing down the overall project. Other phases of the project are critical and would delay the transition to a computer inventory system if they are delayed beyond their expected completion.

Jim Riley's Report

Relying on the information in the PERT network and the computations in Table 16-3, Jim Riley sent the memorandum shown in Figure 16-2 to the company's president.

MEMORANDUM

TO: Clifford Johnson, President, Pharmco Manu-
         facturing Company

FROM: Jim Riley, Manager of Production Control

SUBJECT: Timetable for installation of computerized inven-
         tory control system

The transition to a computer-based system for inventory
management should be completed in approximately 16
months. The following timetable should be followed in
reaching the target date.

Activity	Time within which it should be completed, months
Internal study	2
Analysis of available systems	3
Solicitation of bids	4
Analysis of bids and selection of system	7
Training of personnel who will use system	8
Site preparation	8
Arrival of computer from manufacturer	11
Installation of the computer system	13
Completion of pilot run and preliminary testing	14
Transfer of records and official use of system	16

Figure 16-2   Jim Riley's memorandum.

APPENDIXES

Appendix A

Glossary of Terms

ABC Inventory Classification A system for classifying inventory items according to their contribution to total inventory investment.

Absolute Value The positive value of a number, even if it is preceded by a minus sign. Symbol: | |

Alternate Hypothesis The statement of the condition in a hypothesis test presumed to exist if the null hypothesis is not true.

Analytic Model A mathematical model that is "solved" through the use of standard mathematical or statistical procedures.

Array A set of elements arranged according to ascending or descending order.

Arrival Process The pattern, often probabilistic, in which units enter a queuing system.

Autocorrelation A condition in which a variable is related to its own past performance.

Bar Graph A graphical representation of a frequency distribution in which the length of a bar at the class midpoint represents the class frequency.

Bayesian Statistics The branch of statistics in which empirical or logical probabilities are revised with subjective probabilities.

Binomial (Probability Distribution) A distribution of probabilities in which the random variable is the discrete number of times x an event can be expected to occur out of n trials, given the probability p that it will occur in one trial.

Bivariate Pertaining to two variables; a special case of multivariate; for example, simple linear regression is a form of bivariate analysis.

Central-Limit Theorem The theorem that states that sampling distributions approach a normal distribution as the sample size n increases.

Certainty A decision situation in which the state of nature that will occur is known.

309

Channel Structure   The number of waiting lines and service facilities in a queuing system.

Class   A subset that includes all elements within a specified range, such as the subset of persons aged 20 to 25 years.

Coefficient of Correlation   The square root of the coefficient of determination; a measure of the strength of the regression relationship. Symbols: r (simple regression), R (multiple regression).

Coefficient of Determination   A measure showing the amount, expressed as a percentage, of variation in the dependent variable that is explained by variation in the independent variable(s). Symbols: r^2 (simple regression), R^2 (multiple regression).

Conditional Probability   The probability of an event given that another, specified event has already occurred.

Continuous Data   A set of data elements that can have fractional as well as integer values.

Convenience (Sample)   A sample in which elements are selected on the basis of ease of collection.

Critical Path   The most time-consuming path through a PERT network. The critical path provides an estimate for the duration of the project.

Decile   One of nine points that divide an array into ten subsets of equal numbers of elements. Symbol: D.

Decision Variables   The variables that must be assigned values in a linear-programming model in order to develop a solution to the problem.

Degrees of Freedom   The number of elements in a sample less the number of estimated parameters involved in the associated computation.

Dependent Variable   A variable whose value is determined by other, independent variables. In regression analysis, the variable plotted on the vertical, or Y, axis.

Deterministic Model   A mathematical model in which all the data inputs are assumed to be known with certainty.

Discrete Data   A set of data elements that have only certain specified values, usually integer values.

Economic Order Quantity (EOQ)   The quantity of an inventory item that should be ordered to minimize the sum of the ordering, carrying, and stock-out costs for the item.

Element   A single item within a set. For numerical values, often called a *data* element.

Empirical Probability   The ratio of the number of times a specified event has occurred in the past to the total opportunities for occurrence.

Expected Value   The mean of a probability distribution, given by the sum of each data element times its probability. Symbol: EV.

Feasible-Solution Space   The area of the graph of a linear-programming model containing values for the decision variables that satisfy all the constraints in the problem.

Frame   A listing of all elements in a set from which the sample subset is selected.

Frequency Distribution   A table showing the number of elements in each class of a set.

Histogram A graphical representation of a frequency distribution in which the area of a rectangle represents the class frequency.

Hypothesis Test A form of statistical inference in which a statement concerning a characteristic of a population, usually the value of its mean, is accepted or rejected based upon the value of the corresponding sample characteristic.

Independent Variable A variable whose value is unrelated to values of other variables in the analysis. In regression analysis, the variable plotted on the horizontal, or X, axis.

Intercept The point where an axis is crossed or intersected by another geometric figure, such as a line or a plane.

Interval Estimate The estimated range or interval within which a population parameter can be expected based upon the corresponding sample statistic and the variability of the population or sample.

Joint Probability The probability that two (or more) specified events will both (all) occur.

Judgment (Sample) A sample in which elements are drawn based on the sampler's judgment of their representativeness.

Kurtosis The measure of peakedness in a normal probability distribution curve.

Lead Time The time that elapses between issuing a replenishment order for an inventory item and the time the item is actually received and entered into inventory.

Linear Equation An equation that plots on a graph as a straight line.

Linear Programming A mathematical technique for finding the best use of some set of limited resources.

Linear Regression A form of multivariate analysis that shows the relationship between two or more variables.

Logical Probability The ratio of the number of ways a specified event can occur to the total number of ways all outcomes can occur.

Mean A measure of central tendency, commonly called the *average*. The *arithmetic* mean of a set is the sum of its element divided by the number of elements. Symbols: μ (population), $\overline{X}$ (sample), and others.

Median A measure of central tendency; the central value of an array.

Mode A measure of central tendency; the data element that occurs most frequently within a set.

Model Any abstract representation of reality used to analyze, explain, or predict the real thing it represents.

Monte Carlo Simulation Imitating a phenomenon by repeatedly sampling at random from the probability distribution that describes it.

Multicollinearity A condition in which the relationship between independent variables in a multiple-regression analysis distorts their true relationship to the dependent variable.

Multivariate Pertaining to two or more variables; for example, simple and multiple regression are forms of multivariate analysis.

Network A schematic diagram consisting of arrows and nodes that illustrates the relationships between the activities in a large-scale project.

Normal (Probability Distribution) A distribution of probabilities in which the random variable is continuous and represents the value of a data element in the set, given a mean μ and a standard deviation σ.

Null Hypothesis The statement, usually reflecting the null or no-change condition, about a population characteristic that is tested in a hypothesis test. Symbol: H_0.

Numeric Model A mathematical model that is not solved but experimented with in order to learn the probable outcome of a decision.

Objective Function A mathematical statement of what is to be accomplished by a linear-programming model.

Ogive A curve formed by connecting graphical plots of cumulative or decumulative frequency distributions.

Operating-Characteristics Curve A graph showing the probability of making a Type II error given various values of the true population mean.

Optimization Model A mathematical model designed to find the "best" solution to a problem.

Parameter A characteristic which by itself or with other such characteristics uniquely defines a population.

Percentile One of 99 points that divide an array into 100 subsets of equal numbers of elements. Symbol: P.

PERT (Performance Evaluation and Review Technique) A scheduling technique for large-scale projects based on a network diagram of the activities to be performed.

Point Estimate A single-value estimate, usually of a population parameter based upon the value of the corresponding sample statistic.

Poisson (Probability Distribution) A distribution of probabilities in which the random variable is the discrete number of times (X) an event can be expected to occur given the mean number of occurrences (λ) over some period of time or other measure.

Population The set that contains all data elements relevant to a statistical analysis.

Power Curve A graph showing the probability of making a Type I error for various values of the true population mean.

Probabilistic Model A mathematical model in which one or more of the variables is uncertain and must be represented by a probability distribution.

Probability The likelihood that an event will occur; the number of times (as a percentage) a specified event is expected to occur out of a number of trials or opportunities for occurrence.

Probability Distribution A table or graph that shows the probability of occurrence for each element within a set.

Process Generator An artificial mechanism used in a simulation model to duplicate or imitate the occurrence of a variable of interest.

Quartile One of three points that divide an array into four subsets with equal numbers of elements. Symbol: Q.

Queue Discipline The priority rule that governs the order in which waiting units are served in a queuing system.

Queuing (Waiting-Line) System A system in which arrivals come to a service facility and a waiting line forms if the facility is busy.

Quota (Sample)   A sample in which assigned numbers (quotas) of elements are selected judgmentally from specified strata.

Random (Sample)   A sample in which elements are selected randomly; that is, every element in the set has an equal probability of selection.

Random Variable   A variable whose numerical value is the function of a probabilistic process.

Range   A measure of variability in a set given by the difference between the largest and smallest elements.

Reciprocal   The value of a number when it is divided into 1. The reciprocal of 3 is $\frac{1}{3}$.

Reorder Point   The point at which a replenishment order for an inventory item should be placed (expressed in terms of the number of items remaining in stock).

Risk   A decision situation in which probabilities are used to express the likelihood that each possible state of nature will occur.

Root   The number which when raised to an exponential power gives a specified value. For example, the third (or cube) root of 8 is 2 because $2^3 = 8$.

Safety Stock   The difference between the reorder point and the expected use of an inventory item during its lead time.

Sample   A subset of data elements selected from a population, usually for the purpose of inferring some population characteristic.

Sampling Distribution   A probability distribution of the values of a sample statistic, usually the mean, for all possible samples of a given size n that can be drawn from a population.

Sensitivity Analysis   A procedure for determining how the solution to a model would be affected by a change in one or more of the assumptions or conditions under which the model was built.

Service Process   The time, often expressed in the form of a probability distribution, required to serve arrivals in a queuing system.

Set   A collection of related elements. Symbol: { }.

Significance Level   The risk, expressed as a precentage, that the null hypothesis in a hypothesis test will be rejected when it is true. Symbol: α.

Simplex Method   A systematic procedure for finding the optimal solution to a linear-programming problem.

Simulation   An imitation of some phenomenon in order to explain or predict its behavior.

Skewness   The lack of symmetry in a frequency curve. Positive or right skewness is reflected in an elongated right-hand tail.

Slack   The delay that can be tolerated in completing an activity (or reaching an event) without delaying the scheduled completion of a project.

Slack Variable   A variable that is *added* to a less-than-or-equal-to constraint in a linear-programming model in order to convert it into an equation.

Slope   The rate of change of a line with respect to graphical axes; in a linear equation represented by the letter b and calculated as the change in the Y coordinate per unit increase in the X coordinate.

Standard Deviation   A measure of variability in a data set that considers every element in the set. Symbols: σ (population), s (sample), and others.

Standard Error of the Estimate A measure of variability of the dependent variable in multivariate data. Symbols: $s_{Y|X}$ (simple regression), $s_{Y|X_1, X_2}$ (multiple regression, two independent variables).

Standard Error of the Mean A measure of variability in a sampling distribution. Symbols: $\sigma_{\bar{X}}$ (population standard deviation known), $s_{\bar{X}}$ (population standard deviation estimated).

Standard Error of the Proportion A measure of variability in the distribution of sample proportions. Symbol: σ_π.

Standard Normal Probability Distribution A normal distribution with a mean of zero and a standard deviation of 1.

States of Nature The possible events that could occur after a decision is made that will ultimately determine how successful the decision is.

Statistic (1) The characteristic of a sample which in a population is called a parameter. (2) A data element (common usage).

Statistical Inference The act of drawing a conclusion about a population based upon the statistical characteristics of a sample.

Statistics (1) The branch of mathematics concerned with probabilistic relationships within and between data sets. (2) A collection of data elements (common usage).

Stratified (Sample) A sample in which representation of certain classes (strata) is assured by the selection technique.

Student's Distribution The t distribution, so named after the pen name of W. M. Gosset, the statistician who first described the sampling errors of small samples.

Subjective Probability An estimated probability.

Subset A set within a set, that is, a number of elements related to one another in a way not shared by the other related elements.

Surplus Variable A variable that is *subtracted* from a greater-than-or-equal-to constraint in a linear-programming model in order to convert it into an equation.

Systematic (Sample) A sample in which elements are drawn according to a regular interval or position within the frame.

t Distribution A sampling distribution of the means of small samples drawn from a normal population. By the Central-Limit Theorem, the t distribution approaches a normal distribution as sample size increases.

Transportation Method A systematic procedure for solving linear-programming problems that seek to achieve the minimum cost distribution of items from a set of sources to a set of destinations.

Type I Error Rejecting the null hypothesis when it is true. The probability of such an error is α, the level of significance, often referred to as the *producer's risk*.

Type II Error Accepting the null hypothesis when it is false. The probability of such an error is designated β and is often referred to as the *consumer's risk*.

Uncertainty A decision situation in which no information at all is available about the likelihood that the various states of nature will occur.

Univariate Pertaining to a single variable; for example, finding the mean of a data set is a form of univariate analysis.

Variance A measure of variability in a data set given by the square of the standard deviation. Symbols: σ^2 (population), s^2 (sample), and others.

Venn Diagram A graphical representation of a set, usually as a rectangle, with its component subsets, usually as circles inside the rectangle.

Appendix B

Glossary of Symbols

Symbol	Meaning	First Reference
a	Number of elements in classes below the median class	55
	Y intercept	25
α	Level of significance	123
b	Number of elements in the median class	55
	Slope	25
β	Probability of a Type II error	131
	Regression equation coefficient	143
C_K	Cost to carry 1 unit of an item in inventory for a specified time (usually 1 year)	210
C_O	Cost to place a replenishment order for an inventory item	210
C_S	Cost of stock-out of an inventory item	221
CI	Class interval	55
CM	Class mark	53
CV	Critical value	123
D	Decile	63
	Annual demand for (or use of) an inventory item	210
E	Maximum allowable error	121
$EV(\ )$	Expected value (of a random variable)	76
f	Frequency (of a class)	37
H_a	Alternate hypothesis	122
H_0	Null hypothesis	122
L	Lower boundary of a class	55
λ	Mean of a Poisson process	82
	Average arrival rate per time period in a queuing system	232

Symbol	Meaning	First Reference
m	Most likely estimate for the completion time of an activity in a project	193
μ	Mean of a population	49
	Mean of a normal probability distribution	85
	Average service rate per time period in a queuing system	232
μ_B	Mean of a binomial distribution	81
μ_P	Mean of a Poisson distribution	84
$\mu_{\bar{x}}$	Mean of a sampling distribution	103
N	Number of elements in a population	36
	Number of orders to be placed for an inventory item per year	221
n	Number of trials in a binomial process	78
	Sample size	100
o	Optimistic estimate for the completion time of an activity in a project	193
P	Percentile	63
	Probability of a stock-out for an inventory item	221
$P(\ )$	Probability (of an event)	68
p	Probability of an event for a single trial in a binomial process	78
	Sample proportion	92
	Pessimistic estimate for the completion time of an activity in a project	193
	Purchase price of an item to be held in inventory	210
π	Population proportion	92
Q	Quartile	62
	Quantity of an inventory item to be ordered at each replenishment cycle (Q_0 indicates optimal order quantity)	210
R	Coefficient of multiple correlation	155
	Reorder point for an inventory item	249
R^2	Coefficient of multiple determination	155
r	Coefficient of simple correlation	147
r^2	Coefficient of simple determination	147
S	Slack or surplus variable in a linear-programming model (sometimes used with a subscript)	180
	Safety stock level of an inventory item	221
	Average number of units in a queuing system	232
s	Standard deviation of a sample	101
$s_{\bar{x}}$	Standard error of the mean (from sample standard deviation)	115
s_Y	Standard deviation of the dependent variable Y	138
$s_{Y|X}$	Standard error of the estimate (simple regression)	150
$s_{Y|X_1,X_2}$	Standard error of the estimate (multiple regression, two independent variables)	157

Symbol	Meaning	First Reference
Σ	Summation (of applicable values of a variable)	14
σ	Standard deviation of a population	58
	Standard deviation of a normal probability distribution	85
σ_B	Standard deviation of a binomial distribution	81
σ_P	Standard deviation of a Poisson distribution	84
$\sigma_{\bar{x}}$	Standard error of the mean (from population standard deviation)	103
σ_π	Standard error of the proportion	92
T_E	Earliest time an event could be expected to be reached in a PERT network	197
T_L	Latest time an event can be reached in a PERT network without delaying the entire project	197
T_S	Average time spent in a queuing system per arrival	232
T_W	Average waiting time per arrival in a queuing system	232
t	Distance in standard deviation or errors (t distribution)	109
t_e	Expected completion time for an activity in a project	193
U	Utilization of the service facility in a queuing system	232
W	Average number of waiting units in a queuing system	232
w	Weight of a variable in computing the weighted mean	54
X	A variable	12
	The random variable in the Poisson probability distribution	82
	The independent variable(s) in regression analysis	140
$\bar{X}$	Mean of a sample	100
x	Random variable in a binomial distribution	79
Y	Dependent variable	138
$\bar{Y}$	Mean of the dependent variable	138
z	Distance in standard deviation or errors (normal distribution)	86
$\hat{\ }$	Estimated (used over another symbol)	114
$\mid$	Given (used in conjunction with two or more other symbols, as in $A\mid B$)	15
$\leq$	Less than or equal to	22
$\geq$	More than or equal to	22
$\mid X\mid$	Absolute value (of X)	21

Appendix C

Glossary of Formulas

Number	Formula	Use
(3-1)	$\mu = \dfrac{\Sigma X}{N}$	Mean of population
(3-2)	$\mu = \dfrac{\Sigma f\text{CM}}{N}$	Mean of frequency distribution
(3-3)	$\mu = \dfrac{\Sigma wX}{\Sigma w}$	Weighted mean
(3-4)	$\text{Median} = L + \text{CI}\left(\dfrac{N/2 - a}{b}\right)$	Median of a frequency distribution
(3-5)	$\sigma = \sqrt{\dfrac{\Sigma(X - \mu)^2}{N}}$	Standard deviation of a population
(3-6)	$\sigma = \sqrt{\dfrac{\Sigma X^2 - (\Sigma X)^2/N}{N}}$	Alternate of (3-5)
(3-7)	$\hat{\sigma} = \sqrt{\dfrac{\Sigma f(\text{CM})^2 - [\Sigma f(\text{CM})]^2/N}{N}}$	σ estimated from frequency distribution
(3-8)	$Q_1 = L_1 + \text{CI}\left(\dfrac{N/4 - a_1}{b_1}\right)$	First quartile of a frequency distribution
(3-9)	$Q_2 = L_2 + \text{CI}\left(\dfrac{2N/4 - a_2}{b_2}\right)$	Second quartile of a frequency distribution
(3-10)	$Q_3 = L_3 + \text{CI}\left(\dfrac{3N/4 - a_3}{b_3}\right)$	Third quartile of a frequency distribution
(4-1)	$P(A|B) = \dfrac{P(A)P(B|A)}{P(B)}$	Bayes' law (of probability)
(4-2)	$\text{EV}(X) = \Sigma XP(X)$	Expected value (of a random variable)
(4-3)	$\mu_B = np$	Mean of a binomial distribution
(4-4)	$\sigma_B = \sqrt{np(1 - p)}$	Standard deviation of a binomial distribution

Number	Formula	Use
(4-5)	$\mu_P = \lambda$	Mean of a Poisson distribution
(4-6)	$\sigma_P = \sqrt{\lambda}$	Standard deviation of a Poisson distribution
(4-7)	$z = \dfrac{X - \mu}{\sigma}$	Distance from X to the mean of a normal distribution, measured in standard deviations
	$z = \dfrac{X - \bar{X}}{s}$	Variation of (4-7) based on sample data
	$z = \dfrac{\bar{X} - \mu}{\sigma_{\bar{X}}}$	Variation of (4-7) for a normal sampling distribution
	$t = \dfrac{\bar{X} - \mu}{s_{\bar{X}}}$	Variation of (4-7) for a t-distributed sampling distribution
(4-8)	Mean proportion $= \pi$	Population proportion
(4-9)	$\sigma_\pi = \sqrt{\dfrac{\pi(1 - \pi)}{n}}$	Standard error of the proportion
(4-10)	$z = \dfrac{p - \pi}{\sigma_\pi}$	Variation of (4-7) for a distribution of proportions
(5-1)	$\bar{X} = \dfrac{\Sigma X}{n}$	Mean of a sample
(5-2)	$s = \sqrt{\dfrac{\Sigma(X - \bar{X})^2}{n - 1}}$	Standard deviation of a sample
(5-3)	$s = \sqrt{\dfrac{\Sigma X^2 - (\Sigma X)^2/n}{n - 1}}$	Alternate of (5-2)
(5-4)	$\sigma_{\bar{X}} = \dfrac{\sigma}{\sqrt{n}} \sqrt{\dfrac{N - n}{N - 1}}$	Standard error of the mean corrected for a finite population
(5-5)	$\sigma_{\bar{X}} = \dfrac{\sigma}{\sqrt{n}}$	Uncorrected standard error of the mean
(6-1)	$\bar{X} \pm z\sigma_{\bar{X}}$	Interval estimate of the population mean
(6-2)	$\bar{X} \pm ts_{\bar{X}}$	Alternate of (6-1) when the sampling distribution is t-distributed
(6-4)	$n = \left(\dfrac{z\sigma}{E}\right)^2$	Minimum sample size given α
(6-5)	$n = \left[\dfrac{(z_2 - z_1)\sigma}{\mu_1 - \mu_2}\right]^2$	Minimum sample size given α and β
(7-1)	$Y = a + bX$	The simple regression line
(7-2)	$r^2 = \dfrac{\text{explained variance}}{\text{total variance}}$	The simple coefficient of determination
(7-3)	Total variance $= \dfrac{\Sigma(Y - \bar{Y})^2}{n - 1}$	Variance of Y from the mean

Number	Formula	Use
(7-4)	Explained variance $= \dfrac{\Sigma(\hat{Y} - \overline{Y})^2}{n-2}$	Variance of $\hat{Y}$ from the mean
(7-5)	$r^2 = \dfrac{\Sigma(\hat{Y} - \overline{Y})^2/(n-2)}{\Sigma(Y - \overline{Y})^2/(n-1)}$	Substitution of (7-3) and (7-4) into (7-2)
(7-6)	$r = \pm \sqrt{\dfrac{\Sigma(\hat{Y} - \overline{Y})^2/(n-2)}{\Sigma(Y - \overline{Y})^2/(n-1)}}$	The simple coefficient of correlation
(7-7)	Unexplained variance $= \dfrac{\Sigma(Y - \hat{Y})^2}{n-2}$	Variance of Y from the regression line
(7-8)	$r^2 = 1 - \dfrac{\Sigma(Y - \hat{Y})^2/(n-2)}{\Sigma(Y - \overline{Y})^2/(n-1)}$	Alternate of (7-5)
(7-9)	$r = \pm \sqrt{1 - \dfrac{\Sigma(Y - \hat{Y})^2/(n-2)}{\Sigma(Y - \overline{Y})^2/(n-1)}}$	Alternate of (7-6)
(7-10)	$s_{Y|x} = \sqrt{\dfrac{\Sigma(Y - \hat{Y})^2}{n-2}}$	The standard error of the estimate (simple regression)
(7-11)	$Y = a + b_1 X_1 + b_2 X_2$	The regression plane
(7-12)	$s_{Y \mid x_1, x_2} = \sqrt{\dfrac{\Sigma(Y - \hat{Y})^2}{n-3}}$	The standard error of the estimate (multiple regression, two independent variables)
(9-1)	$t_e = \dfrac{o + 4m + p}{6}$	The expected completion time of an activity in a project
(10-1)	$TC = \dfrac{D}{Q}C_0 + \dfrac{Q}{2}C_K$	Average annual cost of an inventory item (under assumption of constant and known use)
(10-2)	$Q_0 = \sqrt{\dfrac{2DC_0}{C_K}}$	Economic order quantity (EOQ) (under assumption of constant and known use)
(10-3)	$TC = \dfrac{D}{Q}C_0 + \dfrac{Q}{2}C_K + pD$	Average annual cost of an inventory item when its price varies with the quantity ordered and use is constant
(10-4)	$TC = C_S N P + C_K S$	Average annual cost of the *safety stock* for an inventory item when the cost of a stock-out is independent of the exact number of units short
(11-1)	$U = \dfrac{\lambda}{\mu}$	Utilization of the service facility in a queuing system
(11-2)	$W = \dfrac{\lambda^2}{\mu(\mu - \lambda)}$	Average number of units waiting in a queuing system
(11-3)	$S = U + W$	Average number of units in a queuing system
(11-4)	$T_W = \dfrac{\lambda}{\mu(\mu - \lambda)}$	Average waiting time in a queuing system
(11-5)	$T_S = T_W +$ average service time	Average time spent in a queuing system

Appendix D

Binomial Probabilities [1]

n	x	.01	.05	.10	.15	.20	.25	.30	.35	.40	.45	.50
1	0	.9900	.9500	.9000	.8500	.8000	.7500	.7000	.6500	.6000	.5500	.5000
	1	.0100	.0500	.1000	.1500	.2000	.2500	.3000	.3500	.4000	.4500	.5000
2	0	.9801	.9025	.8100	.7225	.6400	.5625	.4900	.4225	.3600	.3025	.2500
	1	.0198	.0950	.1800	.2550	.3200	.3750	.4200	.4550	.4800	.4950	.5000
	2	.0001	.0025	.0100	.0225	.0400	.0625	.0900	.1225	.1600	.2025	.2500
3	0	.9703	.8574	.7290	.6141	.5120	.4219	.3430	.2746	.2160	.1664	.1250
	1	.0294	.1354	.2430	.3251	.3840	.4219	.4410	.4436	.4320	.4084	.3750
	2	.0003	.0071	.0270	.0574	.0960	.1406	.1890	.2389	.2880	.3341	.3750
	3	.0000	.0001	.0010	.0034	.0080	.0156	.0270	.0429	.0640	.0911	.1250
4	0	.9606	.8145	.6561	.5220	.4096	.3164	.2401	.1785	.1296	.0915	.0625
	1	.0388	.1715	.2916	.3685	.4096	.4219	.4116	.3845	.3456	.2995	.2500
	2	.0006	.0135	.0486	.0975	.1536	.2109	.2646	.3105	.3456	.3675	.3750
	3	.0000	.0005	.0036	.0115	.0258	.0469	.0756	.1115	.1536	.2005	.2500
	4	.0000	.0000	.0001	.0005	.0016	.0039	.0081	.0150	.0256	.0410	.0625
5	0	.9510	.7738	.5905	.4437	.3277	.2373	.1681	.1160	.0778	.0503	.0312
	1	.0480	.2036	.3280	.3915	.4096	.3955	.3602	.3124	.2592	.2059	.1562
	2	.0010	.0214	.0729	.1382	.2048	.2637	.3087	.3364	.3456	.3369	.3125
	3	.0000	.0011	.0081	.0244	.0512	.0879	.1323	.1811	.2304	.2757	.3125
	4	.0000	.0000	.0004	.0022	.0064	.0146	.0284	.0488	.0768	.1128	.1562
	5	.0000	.0000	.0000	.0001	.0003	.0010	.0024	.0053	.0102	.0185	.0312
6	0	.9415	.7351	.5314	.3771	.2621	.1780	.1176	.0754	.0467	.0277	.0156
	1	.0571	.2321	.3543	.3993	.3932	.3560	.3025	.2437	.1866	.1359	.0938
	2	.0014	.0305	.0984	.1762	.2458	.2966	.3241	.3280	.3110	.2780	.2344
	3	.0000	.0021	.0146	.0415	.0819	.1318	.1852	.2355	.2765	.3032	.3125
	4	.0000	.0001	.0012	.0055	.0154	.0330	.0595	.0951	.1382	.1861	.2344
	5	.0000	.0000	.0001	.0004	.0015	.0044	.0102	.0205	.0369	.0609	.0938
	6	.0000	.0000	.0000	.0000	.0001	.0002	.0007	.0018	.0041	.0083	.0156
7	0	.9321	.6983	.4783	.3206	.2097	.1335	.0824	.0490	.0280	.0152	.0078
	1	.0659	.2573	.3720	.3960	.3670	.3115	.2471	.1848	.1306	.0872	.0547
	2	.0020	.0406	.1240	.2097	.2753	.3115	.3177	.2985	.2613	.2140	.1641
	3	.0000	.0036	.0230	.0617	.1147	.1730	.2269	.2679	.2903	.2918	.2734
	4	.0000	.0002	.0026	.0109	.0287	.0577	.0972	.1442	.1935	.2388	.2734

[1]From Leonard Kazmier, *Business Statistics*. Copyright © 1976 by McGraw-Hill, Inc. Used with permission of McGraw-Hill Book Company.

n	x	.01	.05	.10	.15	.20	.25	p .30	.35	.40	.45	.50
7	5	.0000	.0000	.0002	.0012	.0043	.0115	.0250	.0466	.0774	.1172	.1641
	6	.0000	.0000	.0000	.0001	.0004	.0013	.0036	.0084	.0172	.0320	.0547
	7	.0000	.0000	.0000	.0000	.0000	.0001	.0002	.0006	.0016	.0037	.0078
8	0	.9227	.6634	.4305	.2725	.1678	.1002	.0576	.0319	.0168	.0084	.0039
	1	.0746	.2793	.3826	.3847	.3355	.2670	.1977	.1373	.0896	.0548	.0312
	2	.0026	.0515	.1488	.2376	.2936	.3115	.2065	.2587	.2090	.1569	.1094
	3	.0001	.0054	.0331	.0839	.1468	.2076	.2541	.2786	.2787	.2568	.2188
	4	.0000	.0004	.0046	.0185	.0459	.0865	.1361	.1875	.2322	.2627	.2734
	5	.0000	.0000	.0004	.0026	.0092	.0231	.0467	.0808	.1239	.1719	.2188
	6	.0000	.0000	.0000	.0002	.0011	.0038	.0100	.0217	.0413	.0403	.1094
	7	.0000	.0000	.0000	.0000	.0001	.0004	.0012	.0033	.0079	.0164	.0312
	8	.0000	.0000	.0000	.0000	.0000	.0000	.0001	.0002	.0007	.0017	.0039
9	0	.9135	.6302	.3874	.2316	.1342	.0751	.0404	.0207	.0101	.0046	.0020
	1	.0830	.2985	.3874	.3679	.3020	.2253	.1556	.1004	.0605	.0339	.0176
	2	.0034	.0629	.1722	.2597	.3020	.3003	.2668	.2162	.1612	.1110	.0703
	3	.0001	.0077	.0446	.1069	.1762	.2336	.2668	.2716	.2508	.2119	.1641
	4	.0000	.0006	.0074	.0283	.0661	.1168	.1715	.2194	.2508	.2600	.2461
	5	.0000	.0000	.0008	.0050	.0165	.0389	.0735	.1181	.1672	.2128	.2461
	6	.0000	.0000	.0001	.0006	.0028	.0087	.0210	.0424	.0743	.1160	.1641
	7	.0000	.0000	.0000	.0000	.0003	.0012	.0039	.0098	.0212	.0407	.0703
	8	.0000	.0000	.0000	.0000	.0000	.0001	.0004	.0013	.0035	.0083	.0176
	9	.0000	.0000	.0000	.0000	.0000	.0000	.0000	.0001	.0003	.0008	.0020
10	0	.9044	.5987	.3487	.1969	.1074	.0563	.0282	.0135	.0060	.0025	.0010
	1	.0914	.3151	.3874	.3474	.2684	.1877	.1211	.0725	.0403	.0207	.0098
	2	.0042	.0746	.1937	.2759	.3020	.2816	.2335	.1757	.1209	.0763	.0439
	3	.0001	.0105	.0574	.1298	.2013	.2503	.2668	.2522	.2150	.1665	.1172
	4	.0000	.0010	.0112	.0401	.0881	.1460	.2001	.2377	.2508	.2384	.2051
	5	.0000	.0001	.0015	.0085	.0264	.0584	.1029	.1536	.2007	.2340	.2461
	6	.0000	.0000	.0001	.0012	.0055	.0162	.0368	.0689	.1115	.1596	.2051
	7	.0000	.0000	.0000	.0001	.0008	.0031	.0090	.0212	.0425	.0746	.1172
	8	.0000	.0000	.0000	.0000	.0001	.0004	.0014	.0043	.0106	.0229	.0439
	9	.0000	.0000	.0000	.0000	.0000	.0000	.0001	.0005	.0016	.0042	.0098
	10	.0000	.0000	.0000	.0000	.0000	.0000	.0000	.0000	.0001	.0003	.0010
11	0	.8953	.5688	.3138	.1673	.0859	.0422	.0198	.0088	.0036	.0014	.0005
	1	.0995	.3293	.3835	.3248	.2362	.1549	.0932	.0518	.0266	.0125	.0054
	2	.0050	.0867	.2131	.2866	.2953	.2581	.1998	.1395	.0887	.0513	.0269
	3	.0002	.0137	.0710	.1517	.2215	.2581	.2568	.2254	.1774	.1259	.0806
	4	.0000	.0010	.0112	.0401	.0881	.1460	.2001	.2377	.2508	.2384	.2051
	5	.0000	.0001	.0025	.0132	.0388	.0803	.1321	.1830	.2207	.2360	.2256
	6	.0000	.0000	.0003	.0023	.0097	.0268	.0566	.0985	.1471	.1931	.2256
	7	.0000	.0000	.0000	.0003	.0017	.0064	.0173	.0379	.0701	.1128	.1611
	8	.0000	.0000	.0000	.0000	.0002	.0011	.0037	.0102	.0234	.0462	.0806
	9	.0000	.0000	.0000	.0000	.0000	.0001	.0005	.0018	.0052	.0126	.0269
	10	.0000	.0000	.0000	.0000	.0000	.0000	.0000	.0002	.0007	.0021	.0054
	11	.0000	.0000	.0000	.0000	.0000	.0000	.0000	.0000	.0000	.0002	.0005
12	0	.8864	.5404	.2824	.1422	.0687	.0317	.0138	.0057	.0022	.0008	.0002
	1	.1074	.3413	.3766	.3012	.2062	.1267	.0712	.0368	.0174	.0075	.0029
	2	.0060	.0988	.2301	.2924	.2835	.2323	.1678	.1088	.0639	.0339	.0161
	3	.0002	.0173	.0852	.1720	.2362	.2581	.2397	.1954	.1419	.0923	.0537
	4	.0000	.0021	.0213	.0683	.1329	.1936	.2311	.2367	.2128	.1700	.1204
	5	.0000	.0002	.0038	.0193	.0532	.1032	.1585	.2039	.2270	.2225	.1934
	6	.0000	.0000	.0005	.0040	.0155	.0401	.0792	.1281	.1766	.2124	.2256
	7	.0000	.0000	.0000	.0006	.0033	.0115	.0291	.0591	.1009	.1489	.1934
	8	.0000	.0000	.0000	.0001	.0005	.0024	.0078	.0199	.0420	.0762	.1208
	9	.0000	.0000	.0000	.0000	.0001	.0004	.0015	.0048	.0125	.0277	.0537

n	x	.01	.05	.10	.15	.20	.25 p	.30	.35	.40	.45	.50
12	10	.0000	.0000	.0000	.0000	.0000	.0000	.0002	.0008	.0025	.0068	.0161
	11	.0000	.0000	.0000	.0000	.0000	.0000	.0000	.0001	.0003	.0010	.0029
	12	.0000	.0000	.0000	.0000	.0000	.0000	.0000	.0000	.0000	.0001	.0002
13	0	.8775	.5133	.2542	.1209	.0550	.0238	.0097	.0037	.0013	.0004	.0001
	1	.1152	.3512	.3672	.2774	.1787	.1029	.0540	.0259	.0113	.0045	.0016
	2	.0070	.1109	.2448	.2937	.2680	.2059	.1388	.0836	.0453	.0220	.0095
	3	.0003	.0214	.0997	.1900	.2457	.2517	.2181	.1651	.1107	.0660	.0349
	4	.0000	.0028	.0277	.0838	.1535	.2097	.2337	.2222	.1845	.1350	.0873
	5	.0000	.0003	.0055	.0266	.0691	.1258	.1803	.2154	.2214	.1989	.1571
	6	.0000	.0000	.0008	.0063	.0230	.0559	.1030	.1546	.1968	.2169	.2095
	7	.0000	.0000	.0001	.0011	.0058	.0186	.0442	.0833	.1312	.1775	.2095
	8	.0000	.0000	.0001	.0001	.0011	.0047	.0142	.0336	.0656	.1089	.1571
	9	.0000	.0000	.0000	.0000	.0001	.0009	.0034	.0101	.0243	.0495	.0873
	10	.0000	.0000	.0000	.0000	.0000	.0001	.0006	.0022	.0065	.0162	.0349
	11	.0000	.0000	.0000	.0000	.0000	.0000	.0001	.0003	.0012	.0036	.0095
	12	.0000	.0000	.0000	.0000	.0000	.0000	.0000	.0000	.0001	.0005	.0016
	13	.0000	.0000	.0000	.0000	.0000	.0000	.0000	.0000	.0000	.0000	.0001
14	0	.8687	.4877	.2288	.1028	.0440	.0178	.0068	.0024	.0008	.0002	.0001
	1	.1229	.3593	.3559	.2539	.1539	.0832	.0407	.0181	.0073	.0027	.0009
	2	.0081	.1229	.2570	.2912	.2501	.1802	.1134	.0634	.0317	.0141	.0056
	3	.0003	.0259	.1142	.2056	.2501	.2402	.1943	.1366	.0845	.0462	.0222
	4	.0000	.0037	.0349	.0998	.1720	.2202	.2290	.2022	.1549	.1040	.0611
	5	.0000	.0004	.0078	.0352	.0860	.1468	.1963	.2178	.2066	.1701	.1222
	6	.0000	.0000	.0013	.0093	.0322	.0734	.1262	.1759	.2066	.2088	.1833
	7	.0000	.0000	.0002	.0019	.0092	.0280	.0618	.1082	.1574	.1952	.2095
	8	.0000	.0000	.0000	.0003	.0020	.0082	.0232	.0510	.0918	.1398	.1833
	9	.0000	.0000	.0000	.0000	.0003	.0018	.0066	.0183	.0408	.0762	.1222
	10	.0000	.0000	.0000	.0000	.0000	.0003	.0014	.0049	.0136	.0312	.0611
	11	.0000	.0000	.0000	.0000	.0000	.0000	.0002	.0010	.0033	.0093	.0222
	12	.0000	.0000	.0000	.0000	.0000	.0000	.0000	.0001	.0005	.0019	.0056
	13	.0000	.0000	.0000	.0000	.0000	.0000	.0000	.0000	.0001	.0002	.0009
	14	.0000	.0000	.0000	.0000	.0000	.0000	.0000	.0000	.0000	.0000	.0001
15	0	.8601	.4633	.2059	.0874	.0352	.0134	.0047	.0016	.0005	.0001	.0000
	1	.1303	.3658	.3432	.2312	.1319	.0668	.0305	.0126	.0047	.0016	.0005
	2	.0092	.1348	.2669	.2856	.2309	.1559	.0916	.0476	.0219	.0090	.0032
	3	.0004	.0307	.1285	.2184	.2501	.2252	.1700	.1110	.0634	.0318	.0139
	4	.0000	.0049	.0428	.1156	.1876	.2252	.2186	.1792	.1268	.0700	.0417
	5	.0000	.4633	.2059	.0874	.0352	.0134	.0047	.0016	.0005	.0001	.0000
	6	.0000	.0000	.0019	.0132	.0430	.0917	.1472	.1906	.2066	.1914	.1527
	7	.0000	.0000	.0003	.0030	.0138	.0393	.0811	.1319	.1771	.2013	.1964
	8	.0000	.0000	.0000	.0005	.0035	.0131	.0348	.0710	.1181	.1647	.1964
	9	.0000	.0000	.0000	.0001	.0007	.0034	.0116	.0298	.0612	.1048	.1527
	10	.0000	.0000	.0000	.0000	.0001	.0007	.0030	.0096	.0245	.0515	.0916
	11	.0000	.0000	.0000	.0000	.0000	.0001	.0006	.0024	.0074	.0191	.0417
	12	.0000	.0000	.0000	.0000	.0000	.0000	.0001	.0004	.0016	.0052	.0139
	13	.0000	.0000	.0000	.0000	.0000	.0000	.0000	.0001	.0003	.0010	.0032
	14	.0000	.0000	.0000	.0000	.0000	.0000	.0000	.0000	.0000	.0001	.0005
	15	.0000	.0000	.0000	.0000	.0000	.0000	.0000	.0000	.0000	.0000	.0000
16	0	.8515	.4401	.1853	.0743	.0281	.0100	.0033	.0010	.0003	.0001	.0000
	1	.1376	.3706	.3294	.2097	.1126	.0535	.0228	.0087	.0030	.0009	.0002
	2	.0104	.1463	.2745	.2775	.2111	.1336	.0732	.0353	.0150	.0056	.0018
	3	.0005	.0359	.1423	.2285	.2463	.2079	.1465	.0888	.0468	.0215	.0085
	4	.0000	.0061	.0514	.1311	.2001	.2252	.2040	.1553	.1014	.0572	.0278
	5	.0000	.0008	.0137	.0555	.1201	.1802	.2099	.2008	.1623	.1123	.0667
	6	.0000	.0001	.0028	.0180	.0550	.1101	.1649	.1982	.1983	.1684	.1222

n	x	.01	.05	.10	.15	.20	.25	p .30	.35	.40	.45	.50
16	7	.0000	.0000	.0004	.0045	.0197	.0524	.1010	.1524	.1889	.1969	.1746
	8	.0000	.0000	.0001	.0009	.0055	.0197	.0487	.0923	.1417	.1812	.1964
	9	.0000	.0000	.0000	.0001	.0012	.0058	.0185	.0442	.0840	.1318	.1746
	10	.0000	.0000	.0000	.0000	.0002	.0014	.0056	.0167	.0392	.0755	.1222
	11	.0000	.0000	.0000	.0000	.0000	.0002	.0013	.0049	.0142	.0337	.0667
	12	.0000	.0000	.0000	.0000	.0000	.0000	.0002	.0011	.0040	.0115	.0278
	13	.0000	.0000	.0000	.0000	.0000	.0000	.0000	.0002	.0008	.0029	.0085
	14	.0000	.0000	.0000	.0000	.0000	.0000	.0000	.0000	.0001	.0005	.0018
	15	.0000	.0000	.0000	.0000	.0000	.0000	.0000	.0000	.0000	.0001	.0002
	16	.0000	.0000	.0000	.0000	.0000	.0000	.0000	.0000	.0000	.0000	.0000
17	0	.8429	.4181	.1668	.0631	.0225	.0075	.0023	.0007	.0002	.0000	.0000
	1	.1447	.3741	.3150	.1893	.0957	.0426	.0169	.0060	.0019	.0005	.0001
	2	.0117	.1575	.2800	.2673	.1914	.1136	.0581	.0260	.0102	.0035	.0010
	3	.0006	.0415	.1556	.2359	.2393	.1893	.1245	.0701	.0341	.0144	.0052
	4	.0000	.0076	.0605	.1457	.2093	.2209	.1868	.1320	.0796	.0411	.0182
	5	.0000	.0010	.0175	.0668	.1361	.1914	.2081	.1849	.1379	.0875	.0472
	6	.0000	.0001	.0039	.0236	.0680	.1276	.1784	.1991	.1839	.1432	.0944
	7	.0000	.0000	.0007	.0065	.0267	.0668	.1201	.1685	.1927	.1841	.1484
	8	.0000	.0000	.0001	.0014	.0084	.0279	.0644	.1134	.1606	.1883	.1855
	9	.0000	.0000	.0000	.0003	.0021	.0093	.0276	.0611	.1070	.1540	.1855
	10	.0000	.0000	.0000	.0000	.0004	.0025	.0095	.0263	.0571	.1008	.1484
	11	.0000	.0000	.0000	.0000	.0001	.0005	.0026	.0090	.0242	.0525	.0944
	12	.0000	.0000	.0000	.0000	.0000	.0001	.0006	.0024	.0081	.0215	.0472
	13	.0000	.0000	.0000	.0000	.0000	.0000	.0001	.0005	.0021	.0068	.0182
	14	.0000	.0000	.0000	.0000	.0000	.0000	.0000	.0001	.0004	.0016	.0052
	15	.0000	.0000	.0000	.0000	.0000	.0000	.0000	.0000	.0001	.0003	.0010
	16	.0000	.0000	.0000	.0000	.0000	.0000	.0000	.0000	.0000	.0000	.0001
	17	.0000	.0000	.0000	.0000	.0000	.0000	.0000	.0000	.0000	.0000	.0000
18	0	.8345	.3972	.1501	.0536	.0180	.0056	.0016	.0004	.0001	.0003	.0010
	1	.1517	.3763	.3002	.1704	.0811	.0338	.0126	.0042	.0012	.0003	.0001
	2	.0130	.1683	.2835	.2556	.1723	.0958	.0458	.0190	.0069	.0022	.0006
	3	.0007	.0473	.1680	.2406	.2297	.1704	.1046	.0547	.0246	.0095	.0001
	4	.0000	.0093	.0700	.1592	.2153	.2130	.1681	.1104	.0614	.0291	.0117
	5	.0000	.0014	.0218	.0787	.1507	.1988	.2017	.1664	.1146	.0666	.0327
	6	.0000	.0002	.0052	.0301	.0816	.1436	.1873	.1941	.1655	.1181	.0708
	7	.0000	.0000	.0010	.0091	.0350	.0820	.1376	.1792	.1892	.1657	.1214
	8	.0000	.0000	.0002	.0022	.0120	.0376	.0811	.1327	.1734	.1864	.1669
	9	.0000	.0000	.0000	.0004	.0033	.0139	.0386	.0794	.1284	.1694	.1855
	10	.0000	.0000	.0000	.0001	.0008	.0042	.0149	.0385	.0771	.1248	.1669
	11	.0000	.0000	.0000	.0000	.0001	.0010	.0046	.0151	.0374	.0742	.1214
	12	.0000	.0000	.0000	.0000	.0000	.0002	.0012	.0047	.0145	.0354	.0708
	13	.0000	.0000	.0000	.0000	.0000	.0000	.0002	.0012	.0045	.0134	.0327
	14	.0000	.0000	.0000	.0000	.0000	.0000	.0000	.0002	.0011	.0039	.0117
	15	.0000	.0000	.0000	.0000	.0000	.0000	.0000	.0000	.0002	.0009	.0031
	16	.0000	.0000	.0000	.0000	.0000	.0000	.0000	.0000	.0000	.0001	.0006
	17	.0000	.0000	.0000	.0000	.0000	.0000	.0000	.0000	.0000	.0000	.0001
	18	.0000	.0000	.0000	.0000	.0000	.0000	.0000	.0000	.0000	.0000	.0000
19	0	.8262	.3774	.1351	.0456	.0144	.0042	.0011	.0003	.0001	.0000	.0000
	1	.1586	.3774	.2852	.1529	.0685	.0268	.0093	.0029	.0008	.0002	.0000
	2	.0144	.1787	.2852	.2428	.1540	.0803	.0358	.0138	.0046	.0013	.0003
	3	.0008	.0533	.1796	.2428	.2182	.1517	.0869	.0422	.0175	.0062	.0018
	4	.0000	.0112	.0798	.1714	.2182	.2023	.1491	.0909	.0467	.0203	.0074

n	x	.01	.05	.10	.15	.20	.25	.30	.35	.40	.45	.50
19	5	.0000	.0018	.0266	.0907	.1636	.2023	.1916	.1468	.0933	.0497	.0222
	6	.0000	.0002	.0069	.0374	.0955	.1574	.1916	.1844	.1451	.0949	.0518
	7	.0000	.0000	.0014	.0122	.0443	.0974	.1525	.1844	.1797	.1443	.0961
	8	.0000	.0000	.0002	.0032	.0166	.0487	.0981	.1489	.1797	.1771	.1442
	9	.0000	.0000	.0000	.0007	.0051	.0198	.0514	.0980	.1464	.1771	.1762
	10	.0000	.0000	.0000	.0001	.0013	.0066	.0220	.0528	.0976	.1449	.1762
	11	.0000	.0000	.0000	.0000	.0003	.0018	.0077	.0233	.0532	.0970	.1442
	12	.0000	.0000	.0000	.0000	.0000	.0004	.0022	.0083	.0237	.0529	.0961
	13	.0000	.0000	.0000	.0000	.0000	.0001	.0005	.0024	.0085	.0233	.0518
	14	.0000	.0000	.0000	.0000	.0000	.0000	.0001	.0006	.0024	.0082	.0222
	15	.0000	.0000	.0000	.0000	.0000	.0000	.0000	.0001	.0005	.0022	.0074
	16	.0000	.0000	.0000	.0000	.0000	.0000	.0000	.0000	.0001	.0005	.0018
	17	.0000	.0000	.0000	.0000	.0000	.0000	.0000	.0000	.0000	.0001	.0003
	18	.0000	.0000	.0000	.0000	.0000	.0000	.0000	.0000	.0000	.0000	.0000
	19	.0000	.0000	.0000	.0000	.0000	.0000	.0000	.0000	.0000	.0000	.0000
20	0	.8179	.3585	.1216	.0388	.0115	.0032	.0008	.0002	.0000	.0000	.0000
	1	.1652	.3774	.2702	.1368	.0576	.0211	.0068	.0020	.0005	.0001	.0000
	2	.0159	.1887	.2852	.2293	.1369	.0669	.0278	.0100	.0031	.0008	.0002
	3	.0010	.0596	.1901	.2428	.2054	.1339	.0716	.0323	.0123	.0040	.0011
	4	.0000	.0133	.0898	.1821	.2182	.1897	.1304	.0738	.0350	.0139	.0046
	5	.0000	.0022	.0319	.1028	.1746	.2023	.1789	.1272	.0746	.0365	.0148
	6	.0000	.0003	.0089	.0454	.1091	.1686	.1916	.1712	.1244	.0746	.0370
	7	.0000	.0000	.0020	.0160	.0545	.1124	.1643	.1844	.1659	.1221	.0739
	8	.0000	.0000	.0004	.0046	.0222	.0609	.1144	.1614	.1797	.1623	.1201
	9	.0000	.0000	.0001	.0011	.0074	.0271	.0654	.1158	.1597	.1771	.1602
	10	.0000	.0000	.0000	.0002	.0020	.0099	.0308	.0686	.1171	.1593	.1762
	11	.0000	.0000	.0000	.0000	.0005	.0030	.0120	.0336	.0710	.1185	.1602
	12	.0000	.0000	.0000	.0000	.0001	.0008	.0039	.0136	.0355	.0727	.1201
	13	.0000	.0000	.0000	.0000	.0000	.0002	.0010	.0045	.0146	.0366	.0739
	14	.0000	.0000	.0000	.0000	.0000	.0000	.0002	.0012	.0049	.0150	.0370
	15	.0000	.0000	.0000	.0000	.0000	.0000	.0000	.0003	.0013	.0049	.0148
	16	.0000	.0000	.0000	.0000	.0000	.0000	.0000	.0000	.0003	.0013	.0046
	17	.0000	.0000	.0000	.0000	.0000	.0000	.0000	.0000	.0000	.0002	.0011
	18	.0000	.0000	.0000	.0000	.0000	.0000	.0000	.0000	.0000	.0000	.0002
	19	.0000	.0000	.0000	.0000	.0000	.0000	.0000	.0000	.0000	.0000	.0000
	20	.0000	.0000	.0000	.0000	.0000	.0000	.0000	.0000	.0000	.0000	.0000
25	0	.7778	.2774	.0718	.0172	.0038	.0008	.0001	.0000	.0000	.0000	.0000
	1	.1964	.3650	.1994	.0759	.0236	.0063	.0014	.0003	.0000	.0000	.0000
	2	.0238	.2305	.2659	.1607	.0708	.0251	.0074	.0018	.0004	.0001	.0000
	3	.0018	.0930	.2265	.2174	.1358	.0641	.0243	.0076	.0019	.0004	.0001
	4	.0001	.0269	.1384	.2110	.1867	.1175	.0572	.0224	.0071	.0018	.0004
	5	.0000	.0060	.0646	.1564	.1960	.1645	.1030	.0506	.0199	.0063	.0016
	6	.0000	.0010	.0239	.0920	.1633	.1828	.1472	.0908	.0442	.0172	.0053
	7	.0000	.0001	.0072	.0441	.1108	.1654	.1712	.1327	.0800	.0381	.0143
	8	.0000	.0000	.0018	.0175	.0623	.1241	.1651	.1607	.1200	.0701	.0322
	9	.0000	.0000	.0004	.0058	.0294	.0781	.1336	.1635	.1511	.1084	.0609
	10	.0000	.0000	.0000	.0016	.0118	.0417	.0916	.1409	.1612	.1419	.0974
	11	.0000	.0000	.0000	.0004	.0040	.0189	.0536	.1034	.1465	.1583	.1328
	12	.0000	.0000	.0000	.0001	.0012	.0074	.0268	.0650	.1140	.1511	.1550
	13	.0000	.0000	.0000	.0000	.0003	.0025	.0115	.0350	.0760	.1236	.1550
	14	.0000	.0000	.0000	.0000	.0000	.0007	.0042	.0161	.0434	.0867	.1328

n	x	.01	.05	.10	.15	.20	.25	p .30	.35	.40	.45	.50
25	15	.0000	.0000	.0000	.0000	.0000	.0002	.0013	.0064	.0212	.0520	.0974
	16	.0000	.0000	.0000	.0000	.0000	.0000	.0004	.0021	.0088	.0266	.0609
	17	.0000	.0000	.0000	.0000	.0000	.0000	.0001	.0006	.0031	.0115	.0322
	18	.0000	.0000	.0000	.0000	.0000	.0000	.0000	.0001	.0009	.0042	.0143
	19	.0000	.0000	.0000	.0000	.0000	.0000	.0000	.0000	.0002	.0013	.0053
	20	.0000	.0000	.0000	.0000	.0000	.0000	.0000	.0000	.0000	.0001	.0016
	21	.0000	.0000	.0000	.0000	.0000	.0000	.0000	.0000	.0000	.0000	.0004
	22	.0000	.0000	.0000	.0000	.0000	.0000	.0000	.0000	.0000	.0000	.0001
30	0	.7397	.2146	.0424	.0076	.0012	.0002	.0000	.0000	.0000	.0000	.0000
	1	.2242	.3389	.1413	.0404	.0093	.0018	.0003	.0000	.0000	.0000	.0000
	2	.0328	.2586	.2277	.1034	.0337	.0086	.0018	.0003	.0000	.0000	.0000
	3	.0031	.1270	.2361	.1703	.0785	.0269	.0072	.0015	.0003	.0000	.0000
	4	.0002	.0451	.1771	.2028	.1325	.0604	.0208	.0056	.0012	.0002	.0000
	5	.0000	.0124	.1023	.1861	.1723	.1047	.0464	.0157	.0041	.0008	.0001
	6	.0000	.0027	.0474	.1368	.1795	.1455	.0829	.0353	.0115	.0029	.0006
	7	.0000.	.0005	.0180	.0828	.1538	.1662	.1219	.0652	.0263	.0081	.0019
	8	.0000	.0001	.0058	.0420	.1106	.1593	.1501	.1009	.0505	.0191	.0055
	9	.0000	.0000	.0016	.0181	.0676	.1298	.1573	.1328	.0823	.0382	.0133
	10	.0000	.0000	.0004	.0067	.0355	.0909	.1416	.1502	.1152	.0656	.0280
	11	.0000	.0000	.0001	.0022	.0161	.0551	.1103	.1471	.1396	.0976	.0509
	12	.0000	.0000	.0000	.0006	.0064	.0291	.0749	.1254	.1474	.1265	.0806
	13	.0000	.0000	.0000	.0001	.0022	.0134	.0444	.0935	.1360	.1433	.1115
	14	.0000	.0000	.0000	.0000	.0007	.0054	.0231	.0611	.1101	.1424	.1354
	15	.0000	.0000	.0000	.0000	.0002	.0019	.0106	.0351	.0783	.1242	.1445
	16	.0000	.0000	.0000	.0000	.0000	.0006	.0042	.0177	.0489	.0953	.1354
	17	.0000	.0000	.0000	.0000	.0000	.0002	.0015	.0079	.0269	.0642	.1115
	18	.0000	.0000	.0000	.0000	.0000	.0000	.0005	.0031	.0129	.0379	.0806
	19	.0000	.0000	.0000	.0000	.0000	.0000	.0001	.0010	.0054	.0196	.0509
	20	.0000	.0000	.0000	.0000	.0000	.0000	.0000	.0003	.0020	.0088	.0280
	21	.0000	.0000	.0000	.0000	.0000	.0000	.0000	.0001	.0006	.0034	.0133
	22	.0000	.0000	.0000	.0000	.0000	.0000	.0000	.0000	.0002	.0012	.0055
	23	.0000	.0000	.0000	.0000	.0000	.0000	.0000	.0000	.0000	.0003	.0019
	24	.0000	.0000	.0000	.0000	.0000	.0000	.0000	.0000	.0000	.0001	.0006
	25	.0000	.0000	.0000	.0000	.0000	.0000	.0000	.0000	.0000	.0000	.0001

Appendix E

Poisson Probabilities [1]

X	λ 0.1	0.2	0.3	0.4	0.5	0.6	0.7	0.8	0.9	1.0
0	.9048	.8187	.7408	.6703	.6065	.5488	.4966	.4493	.4066	.3679
1	.0905	.1637	.2222	.2681	.3033	.3293	.3476	.3595	.3659	.3679
2	.0045	.0164	.0333	.0536	.0758	.0988	.1217	.1438	.1647	.1839
3	.0002	.0011	.0033	.0072	.0126	.0198	.0284	.0383	.0494	.0613
4	.0000	.0001	.0002	.0007	.0016	.0030	.0050	.0077	.0111	.0153
5	.0000	.0000	.0000	.0001	.0002	.0004	.0007	.0012	.0020	.0031
6	.0000	.0000	.0000	.0000	.0000	.0000	.0001	.0002	.0003	.0005
7	.0000	.0000	.0000	.0000	.0000	.0000	.0000	.0000	.0000	.0001

X	λ 1.1	1.2	1.3	1.4	1.5	1.6	1.7	1.8	1.9	2.0
0	.3329	.3012	.2725	.2466	.2231	.2019	.1827	.1653	.1496	.1353
1	.3662	.3614	.3543	.3452	.3347	.3230	.3106	.2975	.2842	.2707
2	.2014	.2169	.2303	.2417	.2510	.2584	.2640	.2678	.2700	.2707
3	.0738	.0867	.0998	.1128	.1255	.1378	.1496	.1607	.1710	.1804
4	.0203	.0260	.0324	.0395	.0471	.0551	.0636	.0723	.0812	.0902
5	.0045	.0062	.0084	.0111	.0141	.0176	.0216	.0260	.0309	.0361
6	.0008	.0012	.0018	.0026	.0035	.0047	.0061	.0078	.0098	.0120
7	.0001	.0002	.0003	.0005	.0008	.0011	.0015	.0020	.0027	.0034
8	.0000	.0000	.0001	.0001	.0001	.0002	.0003	.0005	.0006	.0009
9	.0000	.0000	.0000	.0000	.0000	.0000	.0001	.0001	.0001	.0002

X	λ 2.1	2.2	2.3	2.4	2.5	2.6	2.7	2.8	2.9	3.0
0	.1225	.1108	.1003	.0907	.0821	.0743	.0672	.0608	.0550	.0498
1	.2572	.2438	.2306	.2177	.2052	.1931	.1815	.1703	.1396	.1494
2	.2700	.2681	.2652	.2613	.2565	.2510	.2450	.2384	.2314	.2240
3	.1890	.1966	.2033	.2090	.2138	.2176	.2205	.2225	.2237	.2240
4	.0992	.1082	.1169	.1254	.1336	.1414	.1488	.1557	.1622	.1680

[1]From Leonard Kazmier, *Business Statistics.* Copyright © 1976 by McGraw-Hill, Inc. Used with permission of McGraw-Hill Book Company.

X	2.1	2.2	2.3	2.4	λ 2.5	2.6	2.7	2.8	2.9	3.0
5	.0417	.0476	.0538	.0602	.0668	.0735	.0804	.0872	.0940	.1008
6	.0146	.0174	.0206	.0241	.0278	.0319	.0362	.0407	.0455	.0504
7	.0044	.0055	.0068	.0083	.0099	.0118	.0139	.0163	.0188	.0216
8	.0011	.0015	.0019	.0025	.0031	.0038	.0047	.0057	.0068	.0081
9	.0003	.0004	.0005	.0007	.0009	.0011	.0014	.0018	.0022	.0027
10	.0001	.0001	.0001	.0002	.0002	.0003	.0004	.0005	.0006	.0008
11	.0000	.0000	.0000	.0000	.0000	.0001	.0001	.0001	.0002	.0002
12	.0000	.0000	.0000	.0000	.0000	.0000	.0000	.0000	.0000	.0001

X	3.1	3.2	3.3	3.4	λ 3.5	3.6	3.7	3.8	3.9	4.0
0	.0450	.0408	.0369	.0334	.0302	.0273	.0247	.0224	.0202	.0183
1	.1397	.1304	.1217	.1135	.1057	.0984	.0915	.0850	.0789	.0733
2	.2165	.2087	.2008	.1929	.1850	.1771	.1692	.1615	.1539	.1465
3	.2237	.2226	.2209	.2186	.2158	.2125	.2087	.2046	.2001	.1954
4	.1734	.1781	.1823	.1858	.1888	.1912	.1931	.1944	.1951	.1954
5	.1075	.1140	.1203	.1264	.1322	.1377	.1429	.1477	.1522	.1563
6	.0555	.0608	.0662	.0716	.0771	.0826	.0881	.0936	.0989	.1042
7	.0246	.0278	.0312	.0348	.0385	.0425	.0466	.0508	.0551	.0595
8	.0095	.0111	.0129	.0148	.0169	.0191	.0215	.0241	.0269	.0298
9	.0033	.0040	.0047	.0056	.0066	.0076	.0089	.0102	.0116	.0132
10	.0010	.0013	.0016	.0019	.0023	.0028	.0033	.0039	.0045	.0053
11	.0003	.0004	.0005	.0006	.0007	.0009	.0011	.0013	.0016	.0019
12	.0001	.0001	.0001	.0002	.0002	.0003	.0003	.0004	.0005	.0006
13	.0000	.0000	.0000	.0000	.0001	.0001	.0001	.0001	.0002	.0002
14	.0000	.0000	.0000	.0000	.0000	.0000	.0000	.0000	.0000	.0001

X	4.1	4.2	4.3	4.4	λ 4.5	4.6	4.7	4.8	4.9	5.0
0	.0166	.0150	.0136	.0123	.0111	.0101	.0091	.0082	.0074	.0067
1	.0679	.0630	.0583	.0540	.0500	.0462	.0427	.0395	.0365	.0337
2	.1393	.1323	.1254	.1188	.1125	.1063	.1005	.0948	.0894	.0842
3	.1904	.1852	.1798	.1743	.1687	.1631	.1574	.1517	.1460	.1404
4	.1951	.1944	.1933	.1917	.1898	.1875	.1849	.1820	.1789	.1755
5	.1600	.1633	.1662	.1687	.1708	.1725	.1738	.1747	.1753	.1755
6	.1093	.1143	.1191	.1237	.1281	.1323	.1362	.1398	.1432	.1462
7	.0640	.0686	.0732	.0778	.0824	.0869	.0914	.0959	.1002	.1044
8	.0328	.0360	.0393	.0428	.0463	.0500	.0537	.0575	.0614	.0653
9	.0150	.0168	.0188	.0209	.0232	.0255	.0280	.0307	.0334	.0363
10	.0061	.0071	.0081	.0092	.0104	.0118	.0132	.0147	.0164	.0181
11	.0023	.0027	.0032	.0037	.0043	.0049	.0056	.0064	.0073	.0082
12	.0008	.0009	.0011	.0014	.0016	.0019	.0022	.0026	.0030	.0034
13	.0002	.0003	.0004	.0005	.0006	.0007	.0008	.0009	.0011	.0013
14	.0001	.0001	.0001	.0001	.0002	.0002	.0003	.0003	.0004	.0005
15	.0000	.0000	.0000	.0000	.0001	.0001	.0001	.0001	.0001	.0002

X	5.1	5.2	5.3	5.4	λ 5.5	5.6	5.7	5.8	5.9	6.0
0	.0061	.0055	.0050	.0045	.0041	.0037	.0033	.0030	.0027	.0025
1	.0311	.0287	.0265	.0244	.0225	.0207	.0191	.0176	.0162	.0149
2	.0793	.0746	.0701	.0659	.0618	.0580	.0544	.0509	.0477	.0446
3	.1348	.1293	.1239	.1185	.1133	.1082	.1033	.0985	.0938	.0892
4	.1719	.1681	.1641	.1600	.1558	.1515	.1472	.1428	.1383	.1339
5	.1753	.1748	.1740	.1728	.1714	.1697	.1678	.1656	.1632	.1606
6	.1490	.1515	.1537	.1555	.1571	.1584	.1594	.1601	.1605	.1606
7	.1086	.1125	.1163	.1200	.1234	.1267	.1298	.1326	.1353	.1377
8	.0692	.0731	.0771	.0810	.0849	.0887	.0925	.0962	.0998	.1033
9	.0392	.0423	.0454	.0486	.0519	.0552	.0586	.0620	.0654	.0688
10	.0200	.0220	.0241	.0262	.0285	.0309	.0334	.0359	.0386	.0413
11	.0093	.0104	.0116	.0129	.0143	.0157	.0173	.0190	.0207	.0225
12	.0039	.0045	.0051	.0058	.0065	.0073	.0082	.0092	.0102	.0113
13	.0015	.0018	.0021	.0024	.0028	.0032	.0036	.0041	.0046	.0052
14	.0006	.0007	.0008	.0009	.0011	.0013	.0015	.0017	.0019	.0022
15	.0002	.0002	.0003	.0003	.0004	.0005	.0006	.0007	.0008	.0009
16	.0001	.0001	.0001	.0001	.0001	.0002	.0002	.0002	.0003	.0003
17	.0000	.0000	.0000	.0000	.0000	.0001	.0001	.0001	.0001	.0001

X	6.1	6.2	6.3	6.4	λ 6.5	6.6	6.7	6.8	6.9	7.0
0	.0022	.0020	.0018	.0017	.0015	.0014	.0012	.0011	.0010	.0009
1	.0137	.0126	.0116	.0106	.0098	.0090	.0082	.0076	.0070	.0064
2	.0417	.0390	.0364	.0340	.0318	.0296	.0276	.0258	.0240	.0223
3	.0848	.0806	.0765	.0726	.0688	.0652	.0617	.0584	.0552	.0521
4	.1294	.1249	.1205	.1162	.1118	.1076	.1034	.0992	.0952	.0912
5	.1579	.1549	.1519	.1487	.1454	.1420	.1385	.1349	.1314	.1277
6	.1605	.1601	.1595	.1586	.1575	.1562	.1546	.1529	.1511	.1490
7	.1399	.1418	.1435	.1450	.1462	.1472	.1480	.1486	.1489	.1490
8	.1066	.1099	.1130	.1160	.1188	.1215	.1240	.1263	.1284	.1304
9	.0723	.0757	.0791	.0825	.0858	.0891	.0923	.0954	.0985	.1014
10	.0441	.0469	.0498	.0528	.0558	.0558	.0618	.0649	.0679	.0710
11	.0245	.0265	.0285	.0307	.0330	.0353	.0377	.0401	.0426	.0452
12	.0124	.0137	.0150	.0164	.0179	.0194	.0210	.0227	.0245	.0264
13	.0058	.0065	.0073	.0081	.0089	.0098	.0108	.0119	.0130	.0142
14	.0025	.0029	.0033	.0037	.0041	.0046	.0052	.0058	.0064	.0071
15	.0010	.0012	.0014	.0016	.0018	.0020	.0023	.0026	.0029	.0033
16	.0004	.0005	.0005	.0006	.0007	.0008	.0010	.0011	.0013	.0014
17	.0001	.0002	.0002	.0002	.0003	.0003	.0004	.0004	.0005	.0006
18	.0000	.0001	.0001	.0001	.0001	.0001	.0001	.0002	.0002	.0002
19	.0000	.0000	.0000	.0000	.0000	.0000	.0000	.0001	.0001	.0001

X	λ 7.1	7.2	7.3	7.4	7.5	7.6	7.7	7.8	7.9	8.0
0	.0008	.0007	.0007	.0006	.0006	.0005	.0005	.0004	.0004	.0003
1	.0059	.0054	.0049	.0045	.0041	.0038	.0035	.0032	.0029	.0027
2	.0208	.0194	.0180	.0167	.0156	.0145	.0134	.0125	.0116	.0107
3	.0492	.0464	.0438	.0413	.0389	.0366	.0345	.0324	.0305	.0286
4	.0874	.0836	.0799	.0764	.0729	.0696	.0663	.0632	.0602	.0573
5	.1241	.1204	.1167	.1130	.1094	.1057	.1021	.0986	.0951	.0916
6	.1468	.1445	.1420	.1394	.1367	.1339	.1311	.1282	.1252	.1221
7	.1489	.1486	.1481	.1474	.1465	.1454	.1442	.1428	.1413	.1396
8	.1321	.1337	.1351	.1363	.1373	.1382	.1388	.1392	.1395	.1396
9	.1042	.1070	.1096	.1121	.1144	.1167	.1187	.1207	.1224	.1241
10	.0740	.0770	.0800	.0829	.0858	.0887	.0914	.0941	.0967	.0993
11	.0478	.0504	.0531	.0558	.0585	.0613	.0640	.0667	.0695	.0722
12	.0283	.0303	.0323	.0344	.0366	.0388	.0411	.0434	.0457	.0481
13	.0154	.0168	.0181	.0196	.0211	.0227	.0243	.0260	.0278	.0296
14	.0078	.0086	.0095	.0104	.0113	.0123	.0134	.0145	.0157	.0169
15	.0037	.0041	.0046	.0051	.0057	.0062	.0069	.0075	.0083	.0090
16	.0016	.0019	.0021	.0024	.0026	.0030	.0033	.0037	.0041	.0045
17	.0007	.0008	.0009	.0010	.0012	.0013	.0015	.0017	.0019	.0021
18	.0003	.0003	.0004	.0004	.0005	.0006	.0006	.0007	.0008	.0009
19	.0001	.0001	.0001	.0002	.0002	.0002	.0003	.0003	.0003	.0004
20	.0000	.0000	.0001	.0001	.0001	.0001	.0001	.0001	.0001	.0002
21	.0000	.0000	.0000	.0000	.0000	.0000	.0000	.0000	.0001	.0001

X	λ 8.1	8.2	8.3	8.4	8.5	8.6	8.7	8.8	8.9	9.0
0	.0003	.0003	.0002	.0002	.0002	.0002	.0002	.0002	.0001	.0001
1	.0025	.0023	.0021	.0019	.0017	.0016	.0014	.0013	.0012	.0011
2	.0100	.0092	.0086	.0079	.0074	.0068	.0063	.0058	.0054	.0050
3	.0269	.0252	.0237	.0222	.0208	.0195	.0183	.0171	.0160	.0150
4	.0544	.0517	.0491	.0466	.0443	.0420	.0398	.0377	.0357	.0337
5	.0882	.0849	.0816	.0784	.0752	.0722	.0692	.0663	.0635	.0607
6	.1191	.1160	.1128	.1097	.1066	.1034	.1003	.0972	.0941	.0911
7	.1378	.1358	.1338	.1317	.1294	.1271	.1247	.1222	.1197	.1171
8	.1395	.1392	.1388	.1382	.1375	.1366	.1356	.1344	.1332	.1318
9	.1256	.1269	.1280	.1290	.1299	.1306	.1311	.1315	.1317	.1318
10	.1017	.1040	.1063	.1084	.1104	.1123	.1140	.1157	.1172	.1186
11	.0749	.0776	.0802	.0828	.0853	.0878	.0902	.0925	.0948	.0970
12	.0505	.0530	.0555	.0579	.0604	.0629	.0654	.0679	.0703	.0728
13	.0315	.0334	.0354	.0374	.0395	.0416	.0438	.0459	.0481	.0504
14	.0182	.0196	.0210	.0225	.0240	.0256	.0272	.0289	.0306	.0324

				λ					
X	8.1	8.2	8.3	8.4	8.5	8.6	8.7	8.8	8.9	9.0

X	8.1	8.2	8.3	8.4	8.5	8.6	8.7	8.8	8.9	9.0
15	.0098	.0107	.0116	.0126	.0136	.0147	.0158	.0169	.0182	.0194
16	.0050	.0055	.0060	.0066	.0072	.0079	.0086	.0093	.0101	.0109
17	.0024	.0026	.0029	.0033	.0036	.0040	.0044	.0048	.0053	.0058
18	.0011	.0012	.0014	.0015	.0017	.0019	.0021	.0024	.0026	.0029
19	.0005	.0005	.0006	.0007	.0008	.0009	.0010	.0011	.0012	.0014
20	.0002	.0002	.0002	.0003	.0003	.0004	.0004	.0005	.0005	.0006
21	.0001	.0001	.0001	.0001	.0001	.0002	.0002	.0002	.0002	.0003
22	.0000	.0000	.0000	.0000	.0001	.0001	.0001	.0001	.0001	.0001

				λ					
X	9.1	9.2	9.3	9.4	9.5	9.6	9.7	9.8	9.9	10.0

X	9.1	9.2	9.3	9.4	9.5	9.6	9.7	9.8	9.9	10.0
0	.0001	.0001	.0001	.0001	.0001	.0001	.0001	.0001	.0001	.0000
1	.0010	.0009	.0009	.0008	.0007	.0007	.0006	.0005	.0005	.0005
2	.0046	.0043	.0040	.0037	.0034	.0031	.0029	.0027	.0025	.0023
3	.0140	.0131	.0123	.0115	.0107	.0100	.0093	.0087	.0081	.0076
4	.0319	.0302	.0285	.0269	.0254	.0240	.0226	.0213	.0201	.0189
5	.0581	.0555	.0530	.0506	.0483	.0460	.0439	.0418	.0398	.0378
6	.0881	.0851	.0822	.0793	.0764	.0736	.0709	.0682	.0656	.0631
7	.1145	.1118	.1091	.1064	.1037	.1010	.0982	.0955	.0928	.0901
8	.1302	.1286	.1269	.1251	.1232	.1212	.1191	.1170	.1148	.1126
9	.1317	.1315	.1311	.1306	.1300	.1293	.1284	.1274	.1263	.1251
10	.1198	.1210	.1219	.1228	.1235	.1241	.1245	.1249	.1250	.1251
11	.0991	.1012	.1031	.1049	.1067	.1083	.1098	.1112	.1125	.1137
12	.0752	.0776	.0779	.0822	.0844	.0866	.0888	.0908	.0928	.0948
13	.0526	.0549	.0572	.0594	.0617	.0640	.0662	.0685	.0707	.0729
14	.0342	.0361	.0380	.0399	.0419	.0439	.0459	.0479	.0500	.0521
15	.0208	.0221	.0235	.0250	.0265	.0281	.0297	.0313	.0330	.0347
16	.0118	.0127	.0137	.0147	.0157	.0168	.0180	.0192	.0204	.0217
17	.0063	.0069	.0075	.0081	.0088	.0095	.0103	.0111	.0119	.0128
18	.0032	.0035	.0039	.0042	.0046	.0051	.0055	.0060	.0065	.0071
19	.0015	.0017	.0019	.0021	.0023	.0026	.0028	.0031	.0034	.0037
20	.0007	.0008	.0009	.0010	.0011	.0012	.0014	.0015	.0017	.0019
21	.0003	.0003	.0004	.0004	.0005	.0006	.0006	.0007	.0008	.0009
22	.0001	.0001	.0002	.0002	.0002	.0002	.0003	.0003	.0004	.0004
23	.0000	.0001	.0001	.0001	.0001	.0001	.0001	.0001	.0002	.0002
24	.0000	.0000	.0000	.0000	.0000	.0000	.0000	.0001	.0001	.0001

Appendix F

Probabilities for the Standard Normal Distribution[1]

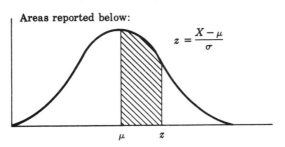

Areas reported below:

$$z = \frac{X - \mu}{\sigma}$$

z	.00	.01	.02	.03	.04	.05	.06	.07	.08	.09
0.0	.0000	.0040	.0080	.0120	.0160	.0199	.0239	.0279	.0319	.0359
0.1	.0398	.0438	.0478	.0517	.0557	.0596	.0636	.0675	.0714	.0753
0.2	.0793	.0832	.0871	.0910	.0948	.0987	.1026	.1064	.1103	.1141
0.3	.1179	.1217	.1255	.1293	.1331	.1368	.1406	.1443	.1480	.1517
0.4	.1554	.1591	.1628	.1664	.1700	.1736	.1772	.1808	.1844	.1879
0.5	.1915	.1950	.1985	.2019	.2054	.2088	.2123	.2157	.2190	.2224
0.6	.2257	.2291	.2324	.2357	.2389	.2422	.2454	.2486	.2518	.2549
0.7	.2580	.2612	.2642	.2673	.2704	.2734	.2764	.2794	.2823	.2852
0.8	.2881	.2910	.2939	.2967	.2995	.3023	.3051	.3078	.3106	.3133
0.9	.3159	.3186	.3212	.3238	.3264	.3289	.3315	.3340	.3365	.3389
1.0	.3413	.3438	.3461	.3485	.3508	.3531	.3554	.3577	.3599	.3621
1.1	.3643	.3665	.3686	.3708	.3729	.3749	.3770	.3790	.3810	.3830
1.2	.3849	.3869	.3888	.3907	.3925	.3944	.3962	.3980	.3997	.4014
1.3	.4032	.4049	.4066	.4082	.4099	.4115	.4131	.4147	.4162	.4177
1.4	.4192	.4207	.4222	.4236	.4251	.4265	.4279	.4292	.4306	.4319

[1]From Leonard Kazmier, *Business Stastics.* Copyright © 1976 by McGraw-Hill, Inc. Used with permission of McGraw-Hill Book Company.

z	.00	.01	.02	.03	.04	.05	.06	.07	.08	.09
1.5	.4332	.4345	.4357	.4370	.4382	.4394	.4406	.4418	.4429	.4441
1.6	.4452	.4463	.4474	.4484	.4495	.4505	.4515	.4525	.4535	.4545
1.7	.4554	.4564	.4573	.4582	.4591	.4599	.4608	.4616	.4625	.4633
1.8	.4641	.4649	.4656	.4664	.4671	.4678	.4686	.4693	.4699	.4706
1.9	.4713	.4719	.4726	.4732	.4738	.4744	.4750	.4756	.4761	.4767
2.0	.4772	.4778	.4783	.4788	.4793	.4798	.4803	.4808	.4812	.4817
2.1	.4821	.4826	.4830	.4834	.4838	.4842	.4846	.4850	.4854	.4857
2.2	.4861	.4864	.4868	.4871	.4875	.4878	.4881	.4884	.4887	.4890
2.3	.4893	.4896	.4898	.4901	.4904	.4906	.4909	.4911	.4913	.4916
2.4	.4918	.4920	.4922	.4925	.4927	.4929	.4931	.4932	.4934	.4936
2.5	.4938	.4940	.4941	.4943	.4945	.4946	.4948	.4949	.4951	.4952
2.6	.4953	.4955	.4956	.4957	.4959	.4960	.4961	.4962	.4963	.4964
2.7	.4965	.4966	.4967	.4968	.4969	.4970	.4971	.4972	.4973	.4974
2.8	.4974	.4975	.4976	.4977	.4977	.4978	.4979	.4979	.4980	.4981
2.9	.4981	.4982	.4983	.4983	.4984	.4984	.4985	.4985	.4986	.4986
3.0	.4987									
3.5	.4997									
4.0	.4999									

Appendix G

t Values for Selected Tail Areas of the t Distributions [1,2]

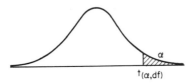

$^{t}(\alpha,df)$

df	$t_{(.10,df)}$	$t_{(.05,df)}$	$t_{(.025,df)}$	$t_{(.01,df)}$	$t_{(.005,df)}$
1	3.078	6.314	12.706	31.821	63.657
2	1.886	2.920	4.303	6.965	9.925
3	1.638	2.353	3.182	4.541	5.841
4	1.533	2.132	2.776	3.747	4.604
5	1.476	2.015	2.571	3.365	4.032
6	1.440	1.943	2.447	3.143	3.707
7	1.415	1.895	2.365	2.998	3.499
8	1.397	1.860	2.306	2.896	3.355
9	1.383	1.833	2.262	2.821	3.250
10	1.372	1.812	2.228	2.764	3.169
11	1.363	1.796	2.201	2.718	3.106
12	1.356	1.782	2.179	2.681	3.055
13	1.350	1.771	2.160	2.650	3.012
14	1.345	1.761	2.145	2.624	2.977
15	1.341	1.753	2.131	2.602	2.947
16	1.337	1.746	2.120	2.583	2.921

[1]Table values are based upon a tail area of α, as in a one-tail hypothesis test. For example, for $\alpha = .05$, df = 15, the value pf t is 1.753. For interval estimation or for a two-tail hypothesis test, tail area is one-half of α. Thus, if $\alpha = .05$ and df = 15, $\alpha/2 = .025$ and the values of t are ±2.131.

[2]Adapted from Odeh, Owen, Birnbaum, and Fisher, *Pocket Book of Statistical Tables*, Marcel Dekker, Inc., New York 1977, pp. 4–5, by permission.

df	$t_{(.10,df)}$	$t_{(.05,df)}$	$t_{(.025,df)}$	$t_{(.01,df)}$	$t_{(.005,df)}$
17	1.333	1.740	2.110	2.567	2.898
18	1.330	1.734	2.101	2.552	2.878
19	1.328	1.729	2.093	2.539	2.861
20	1.325	1.725	2.086	2.528	2.845
21	1.323	1.721	2.080	2.518	2.831
22	1.321	1.717	2.074	2.508	2.819
23	1.319	1.714	2.069	2.500	2.807
24	1.318	1.711	2.064	2.492	2.797
25	1.316	1.708	2.060	2.485	2.787
26	1.315	1.706	2.056	2.479	2.779
27	1.314	1.703	2.052	2.473	2.771
28	1.313	1.701	2.048	2.467	2.763
29	1.311	1.699	2.045	2.462	2.756
∞	1.282	1.645	1.960	2.326	2.576

Appendix H

Random Numbers[1]

Table of Random Numbers

10097	85017	84532	13618	23157	86952	02438	76520
37542	16719	82789	69041	05545	44109	05403	64894
08422	65842	27672	82186	14871	22115	86529	19645
99019	76875	20684	39187	38976	94324	43204	09376
12807	93640	39160	41453	97312	41548	93137	80157
66065	99478	70086	71265	11742	18226	29004	34072
31060	65119	26486	47353	43361	99436	42753	45571
85269	70322	21592	48233	93806	32584	21828	02051
63573	58133	41278	11697	49540	61777	67954	05325
73796	44655	81255	31133	36768	60452	38537	03529
98520	02295	13487	98662	07092	44673	61303	14905
11805	85035	54881	35587	43310	48897	48493	39808
83452	01197	86935	28021	61570	23350	65710	06288
88685	97907	19078	40646	31352	48625	44369	86507
99594	63268	96905	28797	57048	46359	74294	87517
65481	52841	59684	67411	09243	56092	84369	17468
80124	53722	71399	10916	07959	21225	13018	17727
74350	11434	51908	62171	93732	26958	02400	77402
69916	62375	99292	21177	72721	66995	07289	66252
09893	28337	20923	87929	61020	62841	31374	14225
91499	38631	79430	62421	97959	67422	69992	68479
80336	49172	16332	44670	35089	17691	89246	26940
44104	89232	57327	34679	62235	79655	81336	85157
12550	02844	15026	32439	58537	48274	81330	11100
63606	40387	65406	37920	08709	60623	2237	16505

[1]From Leonard Kazimer, *Business Statistics.* Copyright © 1976 by McGraw-Hill, Inc. Used with permission of McGraw-Hill Book Company.

Table of Random Numbers

61196	80240	44177	51171	08723	39323	05798	26457
15474	44910	99321	72173	56239	04595	10836	95270
94557	33663	86347	00926	44915	34823	51770	67897
42481	86430	19102	37420	41976	76559	24358	97344
23523	31379	68588	81675	15694	43438	36879	73208
04493	98086	32533	17767	14523	52494	24826	75246
00549	33185	04805	05431	94598	97654	16232	64051
35963	80951	68953	99634	81949	15307	00406	26898
59808	79752	02529	40200	73742	08391	49140	45427
46058	18633	99970	67348	49329	95236	32537	01390
32179	74029	74717	17674	90446	00597	45240	87379
69234	54178	10805	35635	45266	61406	41941	20117
19565	11664	77602	99817	28573	41430	96382	01758
45155	48324	32135	26803	16213	14938	71961	19476
94864	69074	45753	20505	78317	31994	98145	36168

INDEX

Index